Oxford *Smart*

AQA GCSE PHYSICS

TEACHER HANDBOOK

Jonathan Lansley-Gordon

Editor: Lawrie Ryan

Curriculum Editor: Dr Andrew Chandler-Grevatt

Identity Editor: Jonathan Lansley-Gordon

OXFORD
UNIVERSITY PRESS

Contents

Introduction

About the series

This *Oxford Smart:* AQA GCSE Sciences (9-1) Physics Teacher handbook is part of a blended suite of resources from the UK's number one course for GCSE Science that has been brought up to date to meet the needs of today's science teachers and their students. The Student Book has been approved by AQA and this refreshed edition of the course has been developed in consultation with leading teachers and assessment and cognitive science experts to support all teachers and students to achieve more in their journey through GCSE Science. Student identity is a key focus – examples throughout the course use diverse scientists as well as relevant and contemporary real-world contexts to help students see themselves reflected in the topics, promoting a sense of belonging and equity.

Oxford Smart: AQA GCSE Sciences (9-1) is designed to be flexible, enabling you to co-teach Foundation and Higher tiers, and Trilogy and Separate Sciences. Resources are clearly flagged throughout, helping you to identify the relevant content for your students. Assessments have been developed alongside the resources, ensuring coherence. It is an adaptive course, meaning that there are additional features and functionality on Kerboodle, including automated learning journeys for students and in-depth reporting, which offers data, analysis, and insight. In addition, there is plenty of support for teachers, including practical videos and a CPD and research area with links to current education research.

About your Teacher Handbook

This Teacher Handbook saves you time and effort by offering lesson plans and guidance on differentiation and assessment with a page-by-page match to the Student Book. Based on evidence-informed pedagogies, its clear guidance and explanations support both specialists and non-specialists; included are unit overviews, common misconceptions, tiered terminology, and ideas to help you to relate the content to relevant contexts and students' experiences. Curriculum narrative information is included, so that you can see the sequencing of topics, the preceding knowledge and what comes next.

About the *Oxford Smart* Curriculum for Science

The *Oxford Smart* Curriculum allows for a clear and coherent 11–16 curriculum journey and has been built to address what research shows to be the most pressing considerations for teachers and learners. It connects content, methods, assessment, insight, and evaluation to deliver a holistic learning experience. Drawing on data, insights, and research it equips teachers to deliver personalised learning for all students.

Designed with an explicit egalitarian purpose to enable teachers and students to experience the joy and wonder of learning whilst journeying towards exam readiness and success, the intent of the *Oxford Smart* Curriculum for Science is based on six founding pillars. More information on each pillar can be found in the CPD and Research Hub on Kerboodle.

Six Pillars

Provide clear and coherent pathways

Coherence

The well-researched progression and sequencing support transition from Key Stage 2, through Key Stage 3, and onto GCSE or equivalent qualifications. The pathways are described in the Curriculum narratives documents and embedded throughout Oxford Smart Activate. Unit and chapter openers clearly show the learning journey for students and highlight the common misconceptions identified by our research.

Hold high expectations, aspirations, and ambitions for all learners

High expectations

Oxford Smart Activate identifies strengths, opportunities for challenge, and addresses learning 'gaps'. Expectations are clearly defined using the Developing, Secure, and Extending assessment framework. Challenge and scaffolding are provided in the resources through opt-in, rather than task-based, differentiation.

Enable responsive teaching and learning that continually evolves and improves

Responsive teaching and learning

Regular opportunities for formative assessment are provided through in-book questions, Reactivate questions, mini-checkpoint tests, and checkpoint assessments, with embedded checkpoint follow-up lessons and recommended next steps allowing for easy intervention. Regular spaced practice is included to help students retain concepts.

Support engaged, self-regulated, and metacognitive learning

Metacognitive learning

The Oxford Smart Activate equips learners to succeed with the ability to understand and use learning strategies effectively, facilitating independence and resilience. Learners develop metacognitive skills with scaffolded resources and reflection plenaries for every lesson, while professional development and targeted resources support teachers in modelling metacognitive strategies and self-regulation in the classroom.

Promote development of learner identity and identification with science

Learner identity

Oxford Smart Activate reflects diversity and promotes a sense of belonging for students. The Trigger interest activities and unit openers demonstrate where science impacts our world, enabling learners to make connections with other subjects and their everyday lives.

Stimulate fascination, awe, and wonder in discovery of self and the world around us

Awe and wonder

Oxford Smart Activate promotes the awe and wonder of science through a wide range of hands-on practical activities and hooks for each topic. The unit openers and Trigger interest sections of lessons provide suggestions to spark curiosity and amazement.

Your Teacher Handbook

Unit opener

Introduction to unit
An overview of the unit.

Assessment
Suggested minimum guided learning hours and exam guidance.

Required practicals
This table indicates which required practicals are covered within this unit. It also gives a list of practical skills that could be assessed by that practical.

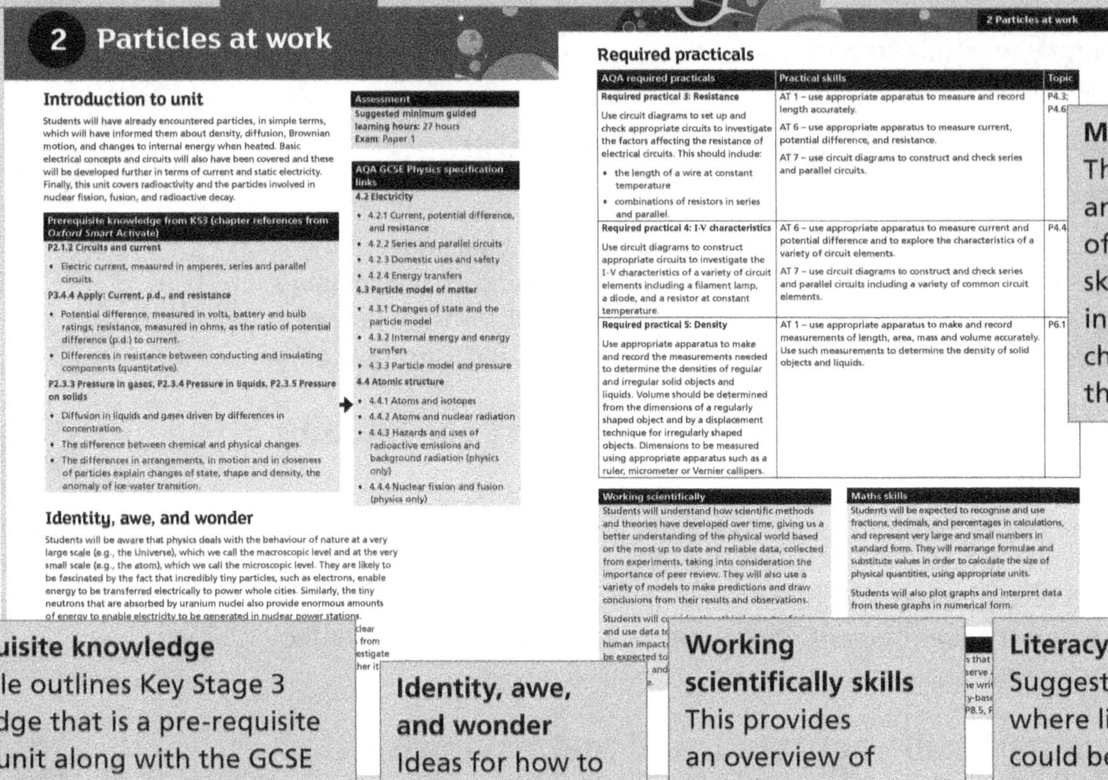

Maths skills
This provides an overview of the maths skills covered in the chapters of the unit.

Pre-requisite knowledge
This table outlines Key Stage 3 knowledge that is a pre-requisite for the unit along with the GCSE specification points that will be covered within the unit. There are also links which cover knowledge from *Oxford Smart* Activate.

Identity, awe, and wonder
Ideas for how to provide context for the unit and make it relevant to students' lives.

Working scientifically skills
This provides an overview of the Working scientifically skills covered in the chapters of the unit.

Literacy skills
Suggestions to where literacy skills could be covered within the unit and a list of topics that include literacy related activities.

Chapter closer

Student reflection
This section suggests how to help students to reflect on their learning over the chapter. Additional strategies to aid reflection can be found in the Metacognition chapter or in the Research hub on Kerboodle.

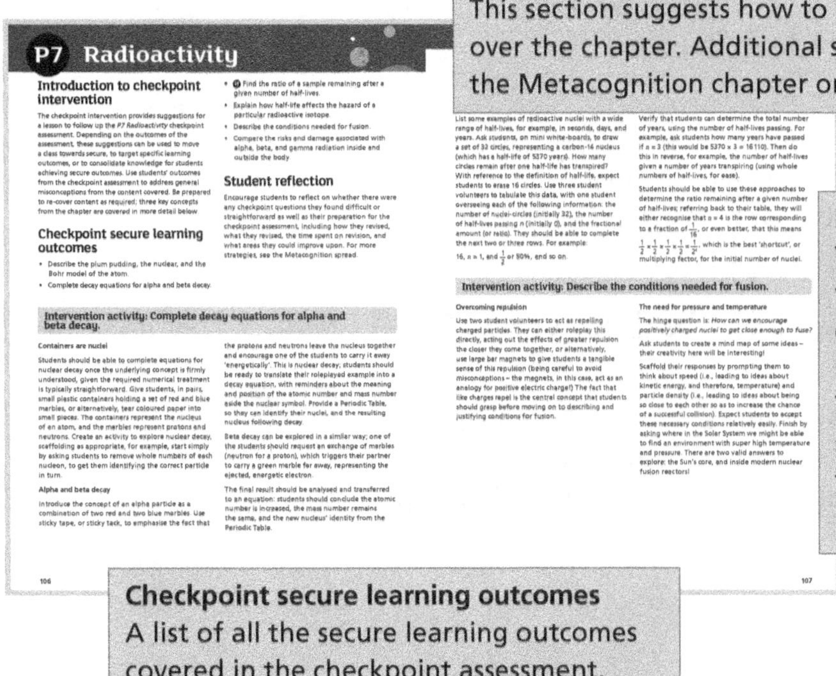

Intervention activities
The intervention activities are suggested follow-up lessons after students have completed the relevant end of chapter Checkpoint Assessment on Kerboodle. The aim of the activities is to move all students from developing to secure on the key learning outcomes covered by the assessment.

Checkpoint secure learning outcomes
A list of all the secure learning outcomes covered in the checkpoint assessment.

Lesson

Suggested lesson plan
A suggested route through the lesson is provided; Trigger Interest, Exposition, Main activity, and Review and practice. The structure has been carefully designed using cognitive science research and is fully supported by resources on Kerboodle including lesson presentations. Suggested timings are given based on a 55-minute lesson.

Core concepts
This indicates the area of the AQA GCSE Physics (9-1) 2016 specification this lesson covers. The theme and sub-theme indicate the area of the *Oxford Smart* Curriculum for Science that this lesson focuses on. Students' understanding can be monitored with the curriculum reports. Relevant Working scientifically links are also provided, as well as the main activity for the lesson.

Physics only

P12 Wave properties

Ⓗ P12.7 Seismic waves

AQA GCSE Physics specification points
Ⓗ 4.6.1.5 Waves for detection and exploration

***Smart Curriculum* Narrative**
Waves

Theme/subtheme
Waves/Seismic waves

Disciplinary knowledge
- WS 1.1, 1.4
- MS 1b

Practical
The seismometer

Learning objectives	Learning outcomes: Developing Aiming for 3	Learning outcomes: Secure Aiming for 5	Learning outcomes: Extending Aiming for 8
Describe the internal structure of the Earth	State that the Earth 's structure consists of the inner core, outer core, mantle, and crust	**Describe the internal structure of the Earth**	Explain the difficulties of exploring inside the Earth
Compare the properties of P and S waves	Describe the properties of P and S waves	**Compare the properties of P and S waves**	Calculate the speed of different types of seismic waves or distance to epicentre using data provided
Describe how P and S waves provide evidence for the structure of the Earth	State that P and S waves provide evidence for the structure of the Earth	**Describe how P and S waves provide evidence for the structure of the Earth**	Explain in detail how to use the detection of seismic waves to determine the structure of the Earth

Tier 2 vocabulary	Tier 3 vocabulary
speed	seismic, seismometer, shadow zone, ultrasound, wave

Lesson plan

Reactivate knowledge

1 What is the difference between longitudinal and transverse waves?
2 Sound waves are longitudinal waves. Which state of matter do they travel fastest in?
3 What happens to waves when they enter another medium and change speed?

Common misconception
Students may think seismic waves travel across the surface of the Earth causing direct damage to buildings. Respond that seismic waves are generated by sudden movement of the Earth's crust, and travel through the core. They are able to shake the surface of the Earth, and the buildings and structures built on it.

Answers: 1 Longitudinal oscillations/vibrations are parallel to direction of energy transfer and in transverse waves they are perpendicular to direction of energy transfer 2 Solid 3 Waves are refracted

Trigger interest

Earthquakes (10 mins)

Set up an activity to explore students' prior knowledge about earthquakes. Create three stations in the classroom, for example, 'real-world examples', 'causes' and 'effects', for student pairs to move around and discuss. Ask targeted questions as appropriate.

Support this activity by providing visual prompts at each station. For example, newspaper articles or short fact sheets to aid conversation.

Exposition

Seismic waves (15 mins)

Outline the structure of the Earth using a large, labelled diagram. Students should be able to identify the different layers, and appreciate that earthquakes move layers of rock, releasing shock waves (seismic waves) that provide information about the Earth's internal structure. Students should contrast this principle to the use of ultrasound waves, which provide information about the internal structures of living organisms.

Discuss the different types of seismic waves and the media they can and cannot travel through. Link the speed of the wave to the time of arrival at the detector. This evidence can be difficult to understand, so ensure the focus is on the meaning of the 'shadow zone', which demonstrates the existence of a liquid outer core.

Main activity

The seismometer (25 mins)

Students investigate the operation of a seismometer by building a simple model. This experiment involves a lot of construction, so ensure roles are split evenly in pairs or amongst groups.

As students complete the practical, circulate groups and discuss simple examples of using seismometer data to calculate the speed of different seismic waves, or examples of how data is used to determine the epicentre of earthquakes.

Review and practice

Make waves (5 mins)

Interactive activity where students choose the missing words to complete a summary on seismic waves and the evidence they provide for the structure of the Earth.

Learning outcomes
This table shows the lesson outcomes which are ramped and divided into three ability bands, Developing (Aiming for Grade 3), Secure (Aiming for Grade 5), and Extending (Aiming for Grade 8).

Vocabulary
Tier 2 vocabulary includes words that are more academic in nature and operate across different contexts. Tier 3 key words are subject-specific. Words in italics have multiple meanings and misconceptions may arise due to this.

Reactivate knowledge
The retrieval questions are included here, along with the answers and any suggestions for discussions relating to any common misconceptions.

Answers to the end-of-spread questions in the Student Book can be found from page 242.

Identity

What is diversity and inclusion?

Diversity, in its simplest form, is different-ness. In the context of curriculum, diversity calls for the presence of different-ness, or variety, in all its forms – a variety of content, people, stories, narratives, perspectives and voices. In seeking a diverse curriculum, the children and young people that we serve are likely to be and feel included – which in turn means they are likely to be engaged and inspired by what they learn. Inclusion, in this sense, is the ideal that all learners, irrespective of their individual identities or backgrounds, are active participants in the curriculum. An inclusive curriculum is integral to a quality education; one that ensures students are equipped to succeed now as well as in their future lives.

The importance of a science identity for GCSE students

A 'science identity' can be thought of as how you see yourself with respect to science. For instance, expressions like 'I could be a scientist', or 'I'm good at science', or even 'I think science is important', are a few of the ways that a strong science identity is articulated.

Students who have strong science identities are likely to:

- be interested in science
- feel competent at science
- be validated by their peers and teachers
- feel like they belong in science.

A strong science identity is a significant predictor of student outcomes as well as the likelihood of persistence with science beyond GCSE. In considering the intent and design of a science curriculum, therefore, it is crucial to consider the ways in which it establishes and reinforces the science identity of all students.

The Identity review process

The Identity review process centred the concept of a science identity; considering content and delivery across each of the sciences and ensuring that young people's diverse personal identities, skills, knowledge and experiences were valued, respected, visible, and represented. Recommendations from best practice in the sector, alongside evidence-based research on the views and opinions of young people doing GCSE science informed our guiding principles, which included:

- Harnessing and valuing individuals' cultural capital and identity, e.g., eliciting and valuing what the student already knows from prior knowledge, news, media, and social media
- Supporting learners to understand how content is relatable and meaningful to themselves at various levels of citizenship – as individuals, communities, and society
- Highlighting scientific elements in broader aspects of everyday life, that students can connect to now as well as in the future (i.e., hobbies, jobs, and careers)
- Considering areas of the curriculum in which diverse knowledge, stories, histories, and experiences are relevant and can be authentically explored
- Considering ways in which both science content as well as the nature of science can be understood by learners

- Supporting all learners to value and behave equitably and inclusively to each other, e.g., encouraging listening, respecting opinions, and ensuring active participation
- Challenging stereotypes about science and scientists by showcasing stories and achievements of people from historically underrepresented backgrounds.

Examples of building science identity in Physics

It was important at the start of the review process to acknowledge that all themes would not be relevant or appropriate to all spreads. This is something we encourage all teachers to consider, both in terms of if/how they choose to implement/tweak the suggestions enclosed, but also in terms of considering the nuances of their own settings and contexts – i.e., the needs of individual students. Alternative activities to approach identity have been provided in the Trigger Interest and Review and practice sections of the lesson so teachers can choose to cover these depending on the needs of the class.

P6.6 Gas pressure and temperature

AQA GCSE Physics specification points
4.3.3.1 Particle motion in gases

***Smart Curriculum* Narrative**
Forces and motion

Theme/subtheme
Forces and motion/Pressure

Disciplinary knowledge
- W.S 1.2, 1.3, 1.4
- MS 4b

Practical
Gas pressure and temperature

Learning objectives	Learning outcomes: Developing Aiming for 3	Learning outcomes: Secure Aiming for 5	Learning outcomes: Extending Aiming for 8
Explain how a gas can exert a pressure	State that gas particles are in constant random motion	**Explain how a gas can exert a pressure**	Explain how the motion of the particles affects the gas pressure
Describe the relationship between an increase in the temperature of a fixed volume of a gas and the increase in pressure of the gas	State that increasing the temperature increases the pressure for a fixed volume of gas	**Describe the relationship between an increase in the temperature of a fixed volume of a gas and the increase in pressure of the gas**	Explain how the linear relationship between temperature and pressure for a fixed volume of gas can be used to determine absolute zero
Describe the behaviour of particles in a gas as the gas is heated	State that the temperature of a gas is related to the kinetic energy of the gas particles	**Describe the behaviour of particles in a gas as the gas is heated**	Explain why gas pressure increases with increasing temperature

Tier 2 vocabulary	Tier 3 vocabulary
density, gas, internal energy, liquid, microscope, molecule, particle, pascal, pressure gauge, random, solid, thermometer, valve	energy transfer, fixed volume, force, kinetic energy, pressure–temperature graph, unit area

Lesson plan

Reactivate knowledge

1 How does the pressure of a gas change if the gas is heated?
2 How do you calculate pressure?
3 How can you increase the pressure acting on a surface?

Common misconception
Students may think gases have no weight (or mass). Emphasise that mass is a measure of the amount of matter. In each state of matter the particles of matter don't vanish, they just have different amounts of internal energy. Imagine a drop of water is placed in a sealed plastic container (with all air excluded). It is then warmed until the droplet turns into water vapour. The total mass stays the same, because the number of particles inside the container stays the same.

Answers: 1 Gas pressure increases 2 Pressure is the force per unit area = force/area 3 Increase the force or decrease the area force acts on

82

P6 Molecules and matter

Trigger interest

Demonstration: Theory of Brownian motion (15 mins)

Show Brownian motion using a smoke cell and discuss the conclusions that can be made using the Working scientifically sheet.

A single smoke cell can be connected to a visualiser so that the whole class can observe it – otherwise, show a video of the observed motion while individuals observe it through the microscope. Compare the observed motion of the smoke particles with pollen grains suspended in water, leading students to conclude that 'invisible' particles suspended in a fluid (i.e., a gas or a liquid!) must exist, because we can observe their effects directly. Keywords and phrases like 'continuous', 'random', and 'collisions' should be provided for additional support where appropriate.

Exposition

Heating gas particles (10 mins)

Discuss the behaviour of gases in terms of particle motion and what happens to particles as their internal energy increases through heating. Ideally, use a gases simulation to demonstrate this in real time, scaffolding questions for students as the gas particles become more energetic.

Expect students to acknowledge collisions with the container walls, and link this to the idea of gas pressure. A balloon is a nice prop to use at this stage – ask for predictions for will happen to the gas pressure if the balloon is heated. Students should explain this in terms of the motion of particles: increased kinetic energy means more collisions (per second) with the balloon walls, until the balloon eventually bursts. Warn any students with sensory differences that there may be a sudden, loud noise. Simulations may also show containers 'bursting' open given enough energy through heating.

Students should be comfortable linking increased temperature to an increased (average) kinetic energy. Extend students further by asking them to consider why the potential energy of the particles is not considered for gases.

Main activity

Demonstration: Gas pressure and temperature (20 mins)

Frame this demonstration as testing the idea that increased temperature means an increased pressure. This gently introduces the students to the idea of proportionality – extend students by providing sample data so that they can plot a graph and describe this proportionality mathematically. They can explore ideas of absolute zero by extrapolating the temperature back to −273 °C.

Demonstrate the heating of a gas as outlined in the practical sheet.

Review and practice

The experimental method and gas laws (10 mins)

Emphasise that Boyle and Charles' experimental work on gases were important in leading to industrial advancements – i.e., understanding the relationship between pressure, volume, and temperature played a crucial role in the development of the steam engine. Students should appreciate that physics plays a crucial role in advancing industry and technology innovations, which are important to any society's economy, health, and prosperity.

This is also an opportunity to touch on the global development and ongoing exchange of scientific knowledge, emphasising that historical scientific progress was not exclusive to Europe and Western scientists: for example, the Islamic scientist Ibn al-Haytham published work on experimental methodology to study light and optics centuries before Boyle and Charles employed similar techniques to study gases. It is thought that his contributions laid the groundwork for the development of the scientific method and had a lasting impact on subsequent scientific thinkers in both Islamic regions and later in Europe.

83

A reflection on the Eurocentric narrative in experimental science by platforming Ibn al-Haytham's work on the experimental method, which is regarded as a likely influence on multiple prominent European scientists centuries later.

Assessment and reporting

About the Curriculum Editor and Assessment Expert

Dr Andrew Chandler-Grevatt has a doctorate in school assessment and a passion for science teaching. Having worked as a teacher for 10 years, this stays at the front of his mind during all of his work in education. Alongside his work with schools and research in school assessment, Andy is a Senior Lecturer in Science education at the University of Brighton and a successful published assessment author.

Welcome from the Curriculum Editor

Welcome to your *Oxford Smart* AQA GCSE Physics Teacher Handbook. The Teacher Handbooks, together with Kerboodle, and the Student Books, provide comprehensive assessment that is coherent with the full 11–16 Oxford Smart Curriculum for Science.

Assessment can be used to set ambitious goals. [It] also plays a key role in enabling responsive teaching and learning needed to support as many learners as possible in reaching that goal. The power of awe and wonder: introducing the new Oxford Smart Curriculum, 2021.

Our assessment is built in across the curriculum and will help your school to monitor progress against a clear framework in order to inform learning, teaching, and assessment from Year 7 to Year 11. Throughout Oxford Smart AQA GCSE for Science, formative assessment has been made easy and we have followed a set of guiding principles.

Oxford Smart AQA GCSE for Science assessment principles

Assessment in *Oxford Smart* AQA GCSE for Science aims to:

- inform teaching and/or learning directly (have a formative function)
- assess agreed and shared objectives that are based on the AQA specifications
- provide opportunities for specific feedback to be given to individual students
- provide opportunities for peer- and self-assessment, when appropriate
- provide insightful data that informs teachers of class and individual progress

A coherent 11–16 learning journey

The *Oxford Smart* Curriculum assessment model is based on bands: the middle band indicates that students have a 'Secure' grasp of the knowledge or skills specified in the curriculum. The band below this is 'Developing' and the band above is 'Extending'. We use bands to help overcome the difficulty of predicting individual grades and should be used to decide on suitable feedback or interventions to help students reach their full potential.

This assessment model has informed the design of the learning outcomes thought the course. Learning outcomes are adapted, and there is a set of learning outcomes for every lesson for each band.

Band	Developing - Aiming for 3			Secure - Aiming for 5			Extending - Aiming for 8		
Grade	1	2	3	4	5	6	7	8	9
Demand	Low			Standard			High		

At GCSE, the framework is based on students' target grades in three bands. This has been informed by how GCSE questions are written at different demands: low, standard, and high demand.

- **Developing/Aiming for 3** targets the students who will have been Developing at KS3 and now have a target grade, aiming for Grade 3 at GCSE (this includes grades 1-3). In this band, students are developing core concepts, and should be able to demonstrate knowledge and understanding and carry out some application and analysis of familiar situations, with some scaffolding. These students will be able to answer most low demand questions and some standard demand questions.
- **Secure/Aiming for 5** is for students working at grades 4 to 6, who would have been secure at KS3. Their knowledge and understanding should be secure and they are able to apply the core concepts to familiar situations. This means that these students should be able to answer low and standard demand GCSE questions.
- **Extending/Aiming for 8** targets those students who have been secure or extending in GCSE and are predicted to get GCSE grade 7 or above. These students have secure knowledge and understanding or all the core concepts and can apply them to more complex or unfamiliar situations. These students will be able to answer GCSE high demand questions.

The checkpoint system

The formative assessment element comprises the checkpoints and mini-checkpoints on Kerboodle. Regular mini-checkpoint assessment tracks progress through the chapter and there is a checkpoint assessment following each chapter. These are all automarked and will help determine if students have a secure understanding of the chapter.

Following each checkpoint or mini-checkpoint, students are offered personalised next steps. They can consolidate their knowledge where they are at a developing level and challenge themselves where they have demonstrated secure knowledge.

Reporting and insights

The formative assessment data will feed online reporting on Kerboodle and live insights into students' strengths and weaknesses to identify misconceptions so you can adapt your teachings.

The *Oxford Smart* Curriculum breaks each science into the Curriculum narrative domains and then splits them into related themes that run through all key stages. The themes are divided into sub-themes to enable targeted interventions.

Kerboodle

Oxford Smart AQA GCSE Sciences (9-1) Kerboodle provides hundreds of engaging lesson resources, as well as a comprehensive assessment package. They are regularly updated, intuitive, and customisable. Resources include lessons and associated resources, automatically marked assessments, books, and the CPD and Research hub.

You can adapt many of the resources to suit your students' needs or your own scheme of work, with all non-interactive activities available as editable documents. You can also upload your existing resources so that everything can be accessed from one location.

Kerboodle is online, allowing you and your students to access the course anytime and anywhere. You can set homework and assessments through the Assessment system and track progress using the Markbook.

Kerboodle Book

The *Oxford Smart* AQA GCSE Sciences Kerboodle Book provides a digital copy of the Student Book for use at the front of the classroom. Teacher access to the Kerboodle Book is automatically available as part of the Lessons, Resources, and Assessment package. You can also purchase additional access for your students. A set of tools is available with the Kerboodle Book so you can personalise your book and make notes. Digital books are hot spotted to direct and relevant access to resources and quizzes on Kerboodle. Like all other resources offered on Kerboodle, the Kerboodle Book can also be accessed using a range of devices.

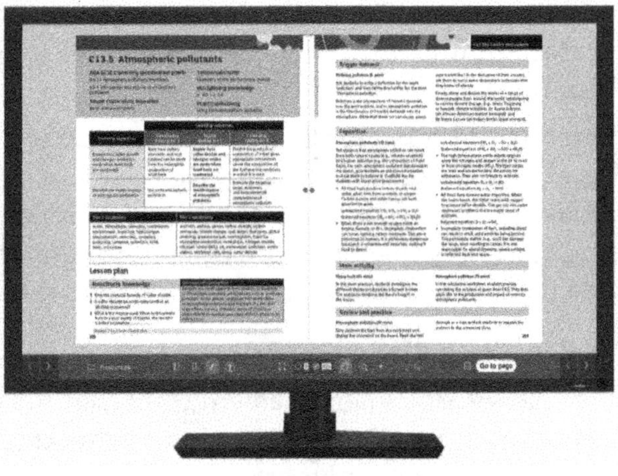

Lessons

Click on the 'Lessons' tab to access the *Oxford Smart* AQA GCSE Sciences Unit planners, Schemes of Work, and Lesson presentations. Each presentation features *resources required*, *Reactivate* question starters, *Something to think about*, lesson objectives, *Trigger Interest* content, activity guidance, key diagrams, questions, extension, plenaries, and homework suggestions. You can further personalise these by adding in your own resources and notes.

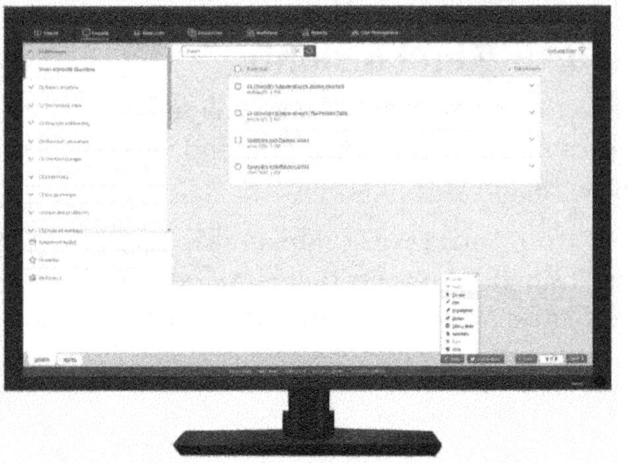

Resources

Click on the 'Resources' tab to access the full list of *Oxford Smart* AQA GCSE Sciences lesson resources.

The resource section includes but are not limited to:

- **Practicals and Activities worksheets** Fully-editable resources to use in class. In addition to an Activity sheet and a Support sheet, which are accessible by pupils, a Teacher and technician sheet provides further ideas on differentiation, answers, and a list of resources required by technicians.
- **Required Practical worksheets** Available in three versions (Aiming for Grades 3/5/8) along with accompanying teacher and technician notes. Every required practical has at least one mini clip to guide students through the methods and data analysis.
- **Maths, Working Scientifically, Literacy, Extension, and Bump up your grade worksheets**
- **Knowledge organisers**
- **Checklists**
- **Glossaries**
- **Metacognition support**

Assessment and Markbook

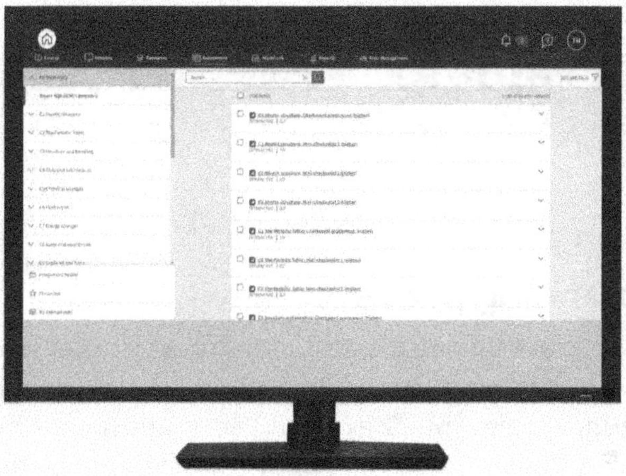

All the assessment material in Kerboodle has been quality assured by our assessment expert. Click on the Assessment tab to find a range of materials to help you deliver a varied, motivating, and effective assessment programme. It's easy to import class registers and create user accounts for your students. Once your classes are set up, you can assign them assessments to do at home, individually, or as a group. A Markbook with reporting function helps you to keep track of your students' results. This includes both automarked assessments and work that needs to be marked by you. Mini-checkpoint and Checkpoint assessments are available for testing during and after a chapter in addition to end of chapter paper-based assessments. Support has been embedded within assessments to motivate students to develop independent study habits. Personalised assessments combined with immediate next steps ensure students understand their results, core concepts are reinforced, and knowledge and skills are stretched by the follow-up questions.

Metacognition

Metacognition is a student's awareness and understanding of their own learning. It is commonly referred to as 'learning about learning' and covers both subject-specific learning as well as wider, independent-study skills. Metacognition can be broken down further into two areas.

Firstly, there is metacognitive knowledge, in which students have a good understanding of the specific tasks they need to complete, as well as the strategies that are available to them with which to carry out the task. Students also have knowledge of themselves as a learner.

Secondly, there is metacognitive regulation, whereby students are able to manage themselves by planning, monitoring, and evaluating their own learning.

An expert learner will naturally follow a cycle like this when they tackle a new task. By explicitly teaching students about this cycle, they will be able to adopt a similar thought process, and this will help them build confidence to successfully complete new and challenging activities.

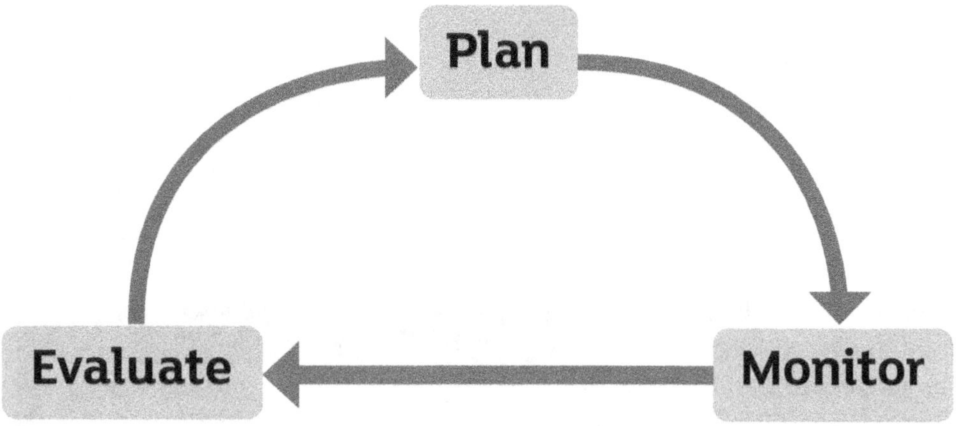

The planning phase

This takes place before the task. An expert learner would plan their approach carefully by thinking about what strategies they could use and what scientific knowledge is relevant. They would also reflect on whether they have completed something similar before and apply this experience to the new task.

The monitoring phase

This takes place during the task. An expert learner would regularly pause and check back to make sure their chosen approach is still right. They would also have the confidence to stop and start again if needed.

The evaluation phase

This takes place at the end of the task. An expert learner would reflect on their performance and carefully consider any areas for improvement. They would plan for how they would use what they have learnt in the future so that they can be successful when doing similar tasks again.

The Seven-Step Approach

The Education Endowment Foundation suggests the following seven-step approach to teaching metacognitive skills.

1 Activate prior knowledge

2 Explicit strategy instruction

3 Modelling of learned strategy

4 Memorisation of strategy

5 Guided practice

6 Independent practice

7 Structured reflection

This approach allows a student to learn about a strategy in relative isolation. When introducing students to a metacognitive strategy, do so in the context of a concept or piece of knowledge that is simple and with which they are already very familiar. This reduces the extra cognitive load that unfamiliar science content would add.

A key aspect of this approach is modelling. Metacognitive modelling is where the focus is on demonstrating how an expert learner would approach a task by verbalising your inner thought processes. For example, you could answer a question 'live' in front of the class. This is helpful for students because they are observing and learning how an expert learner plans, monitors, and evaluates their own learning.

A question tree is a tool that can complement the modelling process. For example, when completing circuit calculations, it is common for values to be placed on a circuit diagram and students then need to choose which values are needed for the calculation. The question tree shown in Figure 1 summarises how students should approach the task.

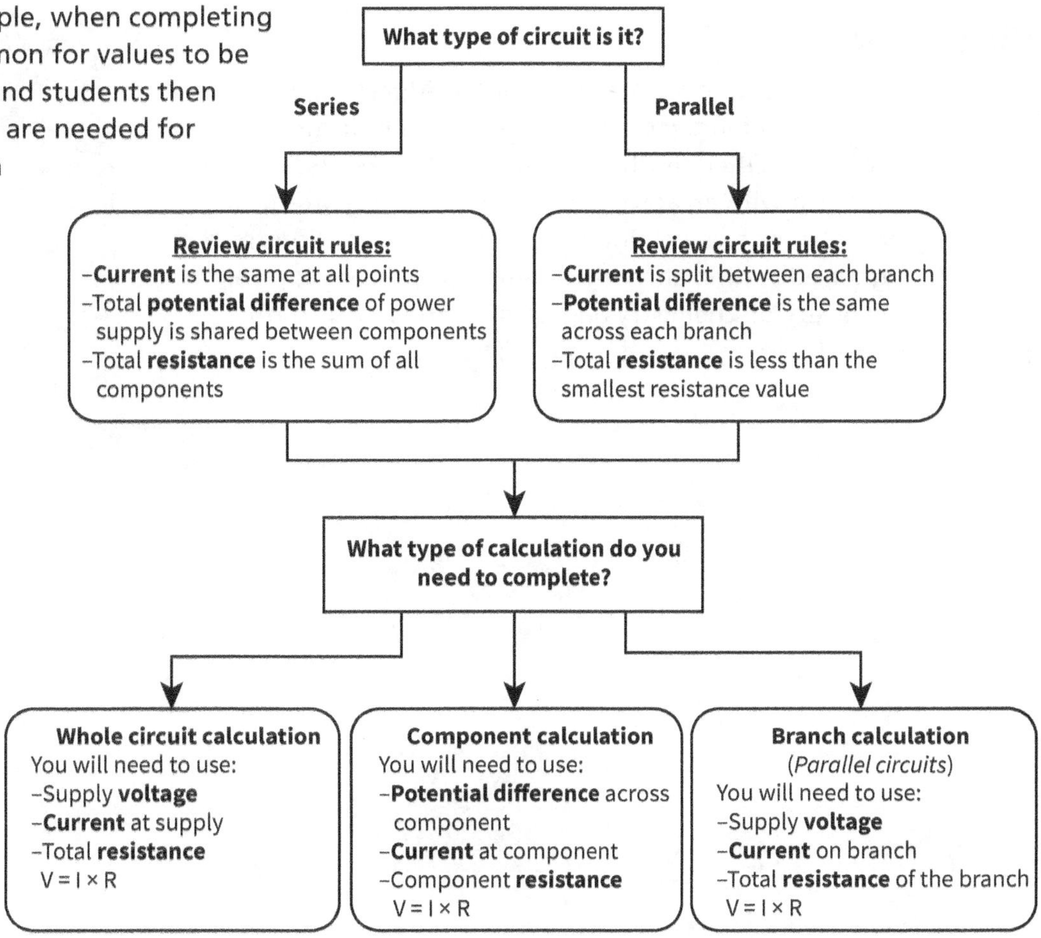

▲ **Figure 1** A question tree for completing circuit calculations

About the authors

Biology

Katie Estruch is an experienced educator with a love of Biology. With 14 years of teaching experience, Katie has held various leadership roles including Head of Science, Head of Biology, and Study Skills Coordinator. In 2019, she took her teaching online too, launching Miss Estruch Biology—an online platform offering biology theory, skills, and revision tips through engaging videos on YouTube, Instagram, and TikTok. Recognising the importance of supporting fellow educators, Katie expanded her reach with the creation of Miss Estruch Teach & Tell. Through this initiative, she shares effective teaching strategies, and lesson plans, and offers support to teachers navigating the ever-evolving educational landscape.

Chemistry

Dr Kristy Turner has a unique role in chemistry education, teaching in both a high school and in the chemistry department at the University of Manchester. A qualified teacher for nearly 20 years, she is a true chemistry specialist with an academic background in synthetic organic chemistry and current research interests including transition from school to higher education and curriculum design. Kristy is a Fellow of the Royal Society of Chemistry and Chair of their education publishing board.

Physics author and Identity Editor

Jonathan Lansley-Gordon obtained his master's degree in theoretical physics from Imperial College London, going on to teach secondary mathematics and science for over 15 years. He is passionate about science communication and believes in the power of physics to inspire, fascinate, and engage all people, regardless of personal characteristics or background. He co-founded The Blackett Lab Family – the first national network of UK based Black physicists – to diversify perceptions of physics and to increase Black representation at all levels in the field. A former Assistant Headteacher and school governor, he now works as a Science Consultant and Director at Physics Forward – an organisation that supports science educators to design and deliver a diverse, inclusive and accessible physics curriculum.

Series Editor

Lawrie Ryan studied Chemistry at University of London before training to be a teacher. He then taught for 20 years, most as Head of Departments, in a variety of large schools in and around Manchester. During this time, he completed a part-time master's degree in science education at Manchester University.

During his career, he has also been a Science Adviser for Norfolk County Council, supporting 50 secondary schools and over 400 primary schools. This involved running many courses for teachers, as well as supporting them in their classrooms with their day-to-day science teaching. During his career he has delivered professional development and worked with teachers in Ireland, Turkey, Barbados, India, Sri Lanka, Lebanon, Jordan, and most of the Gulf states.

He became involved in writing science textbooks whilst teaching in Manchester, and when the opportunity came to pursue this as a full-time career, he became a freelance science education consultant and author/editor. He has been involved in the production of lots of best-selling science resources including Spotlight Science, Chemistry for You (GCSE and A Level editions), the AQA GCSE Sciences series and the Cambridge IGCSE & O Level Essential Science series. As such, he has made science more accessible and enjoyable for students around the world.

1 Energy and energy resources

Introduction to unit

In this unit, students will expand on the concepts relating to energy that they explored during Key Stage 3. Students will already be familiar with energy-related aspects such as the energy content of a variety of foods, and they will also have been exposed to different units used to compare the power ratings and energy transfer of electrical appliances. Fuels and energy resources will also have been covered incorporating the most contemporary issues in relation to the use of renewable and non-renewable energy resources and the quest for a carbon-neutral, net zero environment.

Assessment

Suggested minimum guided learning hours: 17 hours

Exam: Paper 1

Pre-requisite knowledge from KS3 (chapter references from *Oxford Smart* Activate)

P2.2.1 Food and fuels

- Compare energy values of different foods from labels using kJ.

P2.2.8 Energy and power

- Compare power ratings of appliances in W and kW.
- Compare amounts of energy transferred in J, kJ, and kWh.
- Interpret domestic fuel bills and costs.

P3.2.4 Energy resources

- Know a variety of fuels and energy resources.
- Describe a variety of energy changes and transfers.
- Describe and calculate energy changes in systems.

P3.2.5 Apply: Conservation and dissipation

- Know and understand the principle of conservation of energy.

AQA GCSE Physics specification links

4.1 Energy

- 4.1.1 Energy changes in a system, and the ways energy is stored before and after such changes
- 4.1.2 Conservation and dissipation of energy
- 4.1.3 National and global energy resources

Identity, awe, and wonder

Students will be interested in other areas of physics and it would be a good opportunity to get them to consider the importance of the principle of conservation of energy in areas such as forces, waves, radioactivity, and electricity. Regardless of the area, the same principle applies.

Physics has a reputation for being very male-dominated, with many of the theories and discoveries being attributed to men. However, the role of women in physics has been highly significant and, in some cases, largely ignored. It was a woman, Emmy Noether, who was the first person to demonstrate that energy was conserved. This was many years after Aristotle, Newton, and others had suggested that energy is possessed by moving bodies.

Required practicals

AQA required practicals	Practical skills	Topic
Required practical 1: Specific heat capacity An investigation to determine the specific heat capacity of one or more materials. The investigation will involve linking the decrease of one energy store (or work done) to the increase in temperature and subsequent increase in thermal energy stored.	AT 1 – use appropriate apparatus to make and record measurements of mass, time, and temperature accurately. AT 5 – use, in a safe manner, appropriate apparatus to measure energy changes/transfers and associated values such as work done.	P2.2
Required practical 2: Thermal insulation (physics only) Investigate the effectiveness of different materials as thermal insulators and the factors that may affect the thermal insulation properties of a material.	AT 1 – use appropriate apparatus to make and record a range of measurements accurately, including length, area, mass, time, volume, and temperature. AT 5 – use, in a safe manner, appropriate apparatus to measure energy changes/transfers.	P2.1

Working scientifically

Students will be expected to use a variety of models to solve problems, make predictions, and develop scientific explanations in relation to energy stores and transfers and the principle of conservation of energy in physical systems. They will appreciate the power and limitations of science and will consider any ethical issues that may arise in relation to the use of renewable and non-renewable energy resources. Students will become familiar with the correct SI units and most appropriate prefixes when dealing with energy stores, energy transfers, and power (e.g., J, kJ, MW or GW will be used, and converted for use in calculations, as appropriate). Calculations involving correct formulae and units will be executed with answers provided in standard form where appropriate (e.g., 5×10^9 W or 5 GW).

Maths skills

In this topic, students will be expected to analyse data presented as decimals, percentages, or fractions, and in standard form. Students will interpret a variety of charts and tables as well as plot graphs and interpret data in graphical form. They will need to substitute correct values into formulae and provide numerical answers with the correct unit. They will also be expected to change the subject of a formula when necessary, such as when calculating the velocity of a moving object, having been provided with the magnitude of its kinetic energy store.

Literacy links

The practical investigations that students will encounter in this unit will serve as an opportunity for writing skills, such as the writing of a step-by-step method. Other literacy based activities are in P1.5, P1.6, P1.9, P2.1, P2.2, and P3.5.

P1 Conservation and dissipation of energy

P1.1 Changes in energy stores

AQA GCSE Physics specification points

4.1.1.1 Energy stores and systems

Smart Curriculum Narrative

Energy

Theme/subtheme

Energy/Energy stores

Disciplinary knowledge

- WS 1.2, 4.5
- MS 1a, 1c, 2c, 3b, 3c, 4a

Practical

Energy circus

	Learning outcomes		
Learning objectives	**Developing** **Aiming for 3**	**Secure** **Aiming for 5**	**Extending** **Aiming for 8**
Describe a wide range of energy stores in different contexts	State some examples of energy stores	**Describe a wide range of energy stores in different contexts**	Describe the nature of energy stores, in detail, including the relationship between objects
Describe changes in energy stores in terms of the process that causes the change	State the processes that can transfer energy from one store to another	**Describe changes in energy stores in terms of the process that causes the change**	Explain factors that affect the size of changes in energy stores
Describe the changes in how energy is stored in a changing system	Identify changes in some energy stores using simple examples	**Describe the changes in how energy is stored in a changing system**	Represent energy changes graphically, accounting for changes in all stores

Tier 2 vocabulary	Tier 3 vocabulary
energy, force, flow-diagram, heating, mass, temperature, time, waves, *work*	electric current, energy-mass relation, energy store, energy transfer

Lesson plan

Reactivate knowledge

1 Food is an example of which type of energy store?
2 Name one method of transferring energy between stores.
3 Give an example of a kinetic store.

Common misconception

Some students may think that heat is something that moves from place to place. In fact, heat describes the process of the transfer of energy between thermal stores of objects at different temperatures.

Answers: 1 Chemical energy store *2* One from electric current, forces, waves, or heating *3* Any moving object

Trigger interest

Defining energy (5 mins)

Be frank about the fact that energy is a difficult concept to define. Loosely, energy changes are required for things to happen, that is, whilst energy itself is difficult to define, we can observe the effects of energy. In this sense, energy is similar to time!

Exposition

Energy stores and transfers (20 mins)

Introduce the concept of an energy store, the terminology and spelling of the different energy stores and the concept that these can be filled or emptied by energy transfers. It is important that students understand that all objects can contain multiple stores of energy at once, and that these stores can have various amounts of energy at any given moment. Images and videos work as useful tools, with simple prompts to help students decide which energy stores contain energy for a variety of objects.

Flow diagrams

Introduce an example of a flow diagram using a falling bouncy ball. It may not be obvious to students that a force (weight, or the gravitational force) is the process that transfers energy from the ball's gravitational potential store to its kinetic store, as it falls. Use and interchange the phrases 'transfers energy', and 'does work', to model the process.

Students should pick one or two of the demonstrations from the energy circus and draw flow diagrams showing the initial energy store, 'transfer pathway', and final energy store. As an extension, they should consider where there are multiple final energy stores and the transfer mechanisms for these. Encourage discussion and explore ideas; give praise for accurate verbal or written use of the terms 'energy stores' and/or 'transfer pathway' wherever possible.

Main activity

Energy circus (20 mins)

Set up a variety of student-led, explorative demonstrations of different energy stores and transfers. Split students into groups of two or three. Encourage groups to discuss and identify which stores are emptying and which are filling. Scaffold as required, for example, provide a list of types of energy stores. Circulate the groups, asking Socratic questions to open up lines of enquiry and generate ideas.

Once students have had time to identify filling stores and emptying stores for different processes, extend the activity by asking groups to describe how energy from one store is able to be transferred to another. Each group can discuss and present their ideas for a specific station. All students should appreciate that energy is transferred between stores by virtue of a force, an electric current, by waves, or by heating.

Review and practice

Energy misconceptions (10 mins)

As a plenary, examine two or three typical misconceptions around the concept of energy. For example, use a scenario of an errant teacher stating that 'Heat moves from hot objects to cold objects'. Scaffold questions about this scenario by asking, (a) how to use the language of energy stores and transfer pathways to correct this statement and (b) the common misconception about heat being a type of energy store. Emphasise that the process of 'heating' is an energy pathway that allows energy to flow between objects at different temperatures. 'Objects at rest have no energy' is another typical misconception that students can evaluate and improve.

P1.2 Conservation of energy

AQA GCSE Physics specification points

4.1.2.1 Energy transfers in a system

Smart Curriculum Narrative

Energy

Theme/subtheme

Energy/Energy transfer

Disciplinary knowledge

- MS 2c, 4a

Practical

Bungee jumping

	Learning outcomes		
Learning objectives	**Developing Aiming for 3**	**Secure Aiming for 5**	**Extending Aiming for 8**
Apply the law of conservation of energy in straightforward situations	State that energy is conserved in any transfer	**Apply the law of conservation of energy in straightforward situations**	Apply the law of conservation of energy to a wide range of situations and explain where energy has been transferred usefully, stored or dissipated
Describe some examples of closed systems and the changes to energy stores within them	Give an example of a closed system	**Describe some examples of closed systems and the changes to energy stores within them**	Identify a system and explain whether it is closed
Describe the energy changes in a range of experiments and account for energy dissipation to the surroundings	Investigate the energy transfers in a pendulum and a bungee	**Describe the energy changes in a range of experiments and account for energy dissipation to the surroundings**	Evaluate in detail experiments to investigate energy changes

Tier 2 vocabulary	Tier 3 vocabulary
conserved, dissipation, energy, force, heating, magnetic, pendulum, system, *tension*, trajectory, *waste*	closed system, frictional force, gravitational potential energy, kinetic energy, law of conservation of energy, thermal energy

Lesson plan

Reactivate knowledge

1 State the law of conservation of energy.

2 Describe a situation where a kinetic store increases as a gravitational store decreases.

3 Describe a situation where a chemical store decreases and a thermal store increases.

Common misconception

Students may think that energy (usually, thermal energy) is lost to the surroundings. Emphases that all processes (or system changes) dissipate energy. This means it is stored elsewhere, and in less useful ways – typically, the thermal energy stores of the surroundings. This is sometimes referred to as energy being wasted, because it is no longer useable by the original system.

Answers: 1 Energy cannot be created or destroyed *2* Any falling object *3* Burning fuel to heat something

Trigger interest

Rollercoasters (10 mins)

Introduce a video of a rollercoaster and ask if anyone has been on one. Ask if they ever noticed a motor or engine attached to the carriage. Depending on their responses, scaffold follow-up questions to set up the lesson's hinge question.

How does a rollercoaster carriage make it all the way round with no engine or motor to drive it?

By applying the law of the conservation of energy more explicitly to other scenarios throughout the lesson, expect that students should be able to come back to the hinge question and answer it successfully.

Exposition

Forever pendulums (15 mins)

Present the law of conservation of energy, emphasising that students must remember this exactly. Use the law directly to analyse a demonstration of a simple swinging pendulum (e.g., a small mass/bob connected to a string). Recap the energy transfers between gravitational potential stores and kinetic stores, using a labelled diagram to visualise the stores at the maxima and equilibrium points of the pendulum's path. Ensure students understand that this pendulum will not swing forever.

Challenge students to consider where on Earth, or in the Solar System, it would continue to swing for much longer (e.g., on the Moon's surface). Lead students to build up the idea that a closed system is one in which the total energy is always conserved, and to conclude that the simple pendulum swinging on Earth is not a closed system. Students should be able to conclude that frictional forces (i.e., drag, due to Earth's atmosphere) are the energy pathways responsible for transferring energy away from the system to the thermal energy stores of the surroundings.

Main activity

Bungee jumping (20 mins)

Discuss the energy transfers during a bungee jump using a video as a visual prompt – this can be paused at different points to elicit discussion.

Use the experiment to look at the forces involved, gravitational and tension, and reinforce the idea that energy is always conserved in a closed system. Make sure to discuss the heating effect of the bungee rope.

Review and practice

Revisiting rollercoaster (10 mins)

Revisit the hinge question and explore elements of a typical rollercoaster track. Expect some students to be able to conclude that frictional forces between the carriage and the track mean that the roller coaster is not a closed system and the law of conservation of energy means that not all energy is transferred usefully between gravitational potential stores and kinetic stores. For fun, ask students to suggest other means of roller coaster design to minimise dissipation (e.g., magnetic propulsion).

Practice

Tell students that a spacecraft re-entering Earth's atmosphere undergoes a period of rapid deceleration. Challenge students to explain the energy transfers occurring, and determine whether the spacecraft is in a closed or open system. Contrast this with the energy changes involved in a swinging pendulum on the Moon's surface.

P1.3 Energy and work

AQA GCSE Physics specification points

4.5.2 Work done and energy transfer

Smart Curriculum Narrative

Energy

Theme/subtheme

Energy/Energy resources

Disciplinary knowledge

- WS 4.2, 4.3, 4.5
- MS 1c, 3a, 3b, 3c

Practical

Doing work

Learning objectives	Learning outcomes: Developing Aiming for 3	Secure Aiming for 5	Extending Aiming for 8
Describe the action of frictional forces on objects and the associated heating effect	State that energy is measured in joules (J)	**Describe the action of frictional forces on objects and the associated heating effect**	Use the principle of conservation of energy and forces to explain why objects become heated by frictional forces
Use the equation for work done to calculate distances or size of forces	Calculate the work done by a force	**Use the equation for work done to calculate distances or size of forces**	Apply the equation for work done in a wide range of contexts
Use repeat values and calculate a mean to reduce the effect of random errors	Measure the work done by a force experimentally	**Use repeat values and calculate a mean to reduce the effect of random errors**	Evaluate in detail an experiment to measure work done, explaining why there is variation in the measurements

Tier 2 vocabulary	Tier 3 vocabulary
calculation, energy, error, experiment, force, heating, joule, transferred, *work*	energy store, frictional force, work done

Lesson plan

Reactivate knowledge

1 State what work is in science.

2 Describe the factors that affect the amount of work done.

3 State the units of work done.

Common misconception

Students may believe that work describes the amount of 'effort', or force, required to move an object. Remind students that work actually describes the energy transferred to or from an object. This energy transfer happens as a result of an applied force.

Answers: 1 Work is done when a force moves an object *2* The size of the force and the distance moved *3* Joules or Nm

Trigger interest

Defining work (10 mins)

Ask students to define the terms 'work' and 'working' as used in everyday language. Lead them to the idea that physicists define work very carefully, and that it is linked to forces.

Demonstration: Forces and energy

Ask students to describe some situations where forces cause changes in energy stores. Scaffold this using a short demonstration. Drop a tennis ball asking students to decide, from a set of multiple-choice options, which force is responsible for bringing the tennis ball to the ground. Next, they should determine the changes in energy stores, and then conclude that the weight (force due to gravity) is directly responsible for this; it has done work on the tennis ball. Link back to the language about energy stores and pathways introduced in P1.1. Extend this starter by contrasting different sized balls falling from the same height (i.e., varying the working force), or using the same sized balls falling at different heights (i.e., varying the distance moved).

Exposition

Doing work (10 mins)

Perform an example calculation, building on the demonstration of a falling ball. Be deliberate when interchanging the phrases 'work done on the object' and 'the energy transferred', to emphasise the unit of work is that of energy.

Main activity

Doing work (30 mins)

Students carry out the practical, including supporting calculations. There will be considerable errors in the experiments due to frictional forces, which students should discuss.

Discuss the heating effect of frictional forces, using the examples. These can be supported by practical demonstrations such as hand rubbing or bicycle brakes.

Review and practice

Working, or not? (5 mins)

Ask a student volunteer to demonstrate holding a heavy textbook at arm's length, without dropping it. Ask the class, in groups, to discuss and vote as true or false on the proposition that the student does no work on the textbook. Expect groups to anticipate this as true, and create further discussion as to why in fact energy is being transferred to thermal stores as the student holds it. Challenge students to consider what may be happening on a micro-scale as muscle fibres 'work' to keep the book aloft.

Mathematical work out

Students answer additional questions involving the equation for work done in differing contexts. Extension questions include both rearrangement and non-base units, for example, grams.

P1.4 Gravitational potential energy stores

AQA GCSE Physics specification points

4.1.1.1 Energy stores and systems

4.1.1.2 Changes in energy

Smart Curriculum Narrative

Energy

Theme/subtheme

Energy/Energy stores

Disciplinary knowledge

- WS 1.2, 4.3, 4.4, 4.5, 4.6
- MS 1a, 1c, 3a, 3b, 3c

Activity/Practical

GPE calculations/Stepping up

	Learning outcomes		
Learning objectives	**Developing Aiming for 3**	**Secure Aiming for 5**	**Extending Aiming for 8**
Describe the effect of a different gravitational field strength on the gravitational potential energy store changes of a system	State the factors that affect the change in the gravitational potential energy store of a system	**Describe the effect of a different gravitational field strength on the gravitational potential energy store changes of a system**	Apply the gravitational potential energy store equations in a wide range of contexts
Calculate the gravitational potential energy store of a system using the mass, gravitational field strength, and height	Calculate the gravitational potential energy store of a system using the weight of an object and its height	**Calculate the gravitational potential energy store of a system using the mass, gravitational field strength, and height**	Perform calculations using rearrangements of the gravitational potential energy store equations
Identify the measurements needed and determine the change in energy in the gravitational potential store of an object	Measure the gravitational potential energy store changes in a system with a simple practical activity	**Identify the measurements needed and determine the change in energy in the gravitational potential store of an object**	Explain why vertical height is used to measure the change in gravitational potential store for an object going up or down a slope

Tier 2 vocabulary	Tier 3 vocabulary
calculation, distance, energy, equation, force, height, mass, *work*	energy store, gravitational field strength, gravitational potential energy, work done

Lesson plan

Reactivate knowledge

1 State the difference between mass and weight.
2 State the equation to calculate weight.
3 State the equation for work done.

Common misconception

Anticipate students confusing mass and weight, especially if adding 'weight' to a spring, for instance. Distinguish these quantities by emphasising their units; mass in kilograms, kg, and weight in Newtons, N.

Answers: 1 Weight is a force (N) and mass is the amount of matter (kg) *2* Weight = mass × gravitational field strength
3 Work done = force × distance moved in the direction of force

Trigger interest

Lifting work (10 mins)

Remind students of the demonstrations from the previous lesson of work being done by a force when an object moves a distance. Using two student volunteers, with one holding a mass at a steady height and one lifting, ask the class which student does work and to justify their answers by linking to the changes in energy stores. Ask them to identify and explain the factors that affect the size of the energy transfers.

Exposition

Gravitational potential energy transfers (15 mins)

Write the equation for work done in words (work done = force × distance moved in the same direction as the force), then in symbols ($W = Fs$) directly underneath the corresponding words. Ask the students to identify and describe the direct proportionality between the change in distance to work done.

Deliberately erase the word 'distance' and replace with the word 'height', asking students to justify this. Ask for a symbol that might sensibly replace s and lead to the gravitational potential energy equation by arguing that the force required to lift an object must at least equal its weight, mg.

Main activity

Give students the choice of applying the equation experimentally or theoretically. This approach could lead to a well-differentiated lesson with students coming together to share and demonstrate their learning.

GPE calculations (20 mins)

Students should perform some simple calculations of work done and changes in gravitational potential energy stores, including some set on different planets. Extend, as before, by requiring rearrangements to calculate any of mass, gravitational field strength, or height.

Stepping up

Carry out the simple practical experiment to reinforce students' use of the gravitational potential energy equation. This experiment should only take a few minutes; extend by asking students to discuss the sources of uncertainty.

Review and practice

How high? (10 mins)

Ask students to work in groups to calculate the gravitational potential energy of a jumbo jet at cruising altitude. Challenge students to research typical values of the mass of a jumbo jet (400 000 kg) and the cruising altitude (10 km). Explore the consistency of the answers to within orders of magnitude (40 GJ). Support and/or extend students by using standard form and unit prefixes.

P1.5 Kinetic energy and elastic potential energy stores

AQA GCSE Physics specification points

4.1.1.2 Changes in energy

Smart Curriculum Narrative

Energy

Theme/subtheme

Energy/Energy stores

Disciplinary knowledge

- WS 1.2, 4.2, 4.3, 4.6
- MS 1a, 1c, 3a, 3b, 3c

Practical

Investigating kinetic energy stores
Investigating a catapult

Learning objectives	Learning outcomes: Developing Aiming for 3	Secure Aiming for 5	Extending Aiming for 8
Calculate the kinetic energy store of an object	State the factors that affect the size of a kinetic energy store of an object	**Calculate the kinetic energy store of an object**	Perform calculations involving the rearrangement of the kinetic energy equation
Calculate the elastic potential energy store of a stretched spring	State the factors that affect the elastic potential energy store of a spring	**Calculate the elastic potential energy store of a stretched spring**	Perform calculations involving the rearrangement of the elastic potential energy equation
Calculate energy changes involving elastic potential and kinetic energy store	Describe energy changes involving elastic potential energy and kinetic energy stores	**Calculate energy changes involving elastic potential and kinetic energy store**	Perform a wide range of calculations involving transfer of energy

Tier 2 vocabulary	Tier 3 vocabulary
calculation, energy, equation, kilogram (kg), mass, Newton (N), projectile, proton, *store*, transfer, velocity, weight	kinetic energy, elastic potential energy, gravitational potential energy

Lesson plan

Reactivate knowledge

1 What is a kinetic energy store?

2 What is an elastic potential store?

3 State Hooke's law.

Common misconception

Anticipate students confusing mass and weight, especially if adding 'weight' to a spring, for instance. Distinguish these quantities by emphasising that mass is a property of an object while weight is a force acting on the object. They can also be distinguished by their units: mass in kilograms, kg, and weight in Newtons, N.

Answers: 1 Energy an object has when it is moving *2* Energy stored in a springy object when it is stretched or squashed *3* The force is proportional to the extension

Trigger interest

Kinetic cards (5 mins)

Create a card-sorting activity in which students arrange objects, with mass and velocity displayed, in order of increasing kinetic energy stores. Include at least one surprising object, for example, a proton in a particle accelerator with a velocity of 95% of the speed of light. At this stage, this activity relies on their intuition and will be revisited in the plenary. Reassure students they will compare their guesses at the end of the lesson.

Exposition

This lesson introduces two equations and two concepts. Although they are linked, be aware that students will need them to be carefully distinguished and explained before the ideas are unified.

Kinetic energy vs. elastic potential energy (15 mins)

Use two white-boards (or divide a screen in two) to compare and contrast the equation for kinetic energy and elastic potential energy. An image demonstrating each, for example, a fast-moving ball bearing and a large coiled spring, will help students to visualise and distinguish each concept. When appropriate (now or later in the lesson), bring the two objects together visually (e.g., a video of a pinball spring-mechanism) as a way of showing the energy transfer explicitly.

Demonstrate an example calculation for each equation before expecting students to try two or three of their own.

Main activity

Investigating kinetic energy stores (25 mins)

Students carry out the practical to investigate the factors that affect the amount of energy in a kinetic store. They may need reminding beforehand about gravitational potential energy stores.

Investigating a catapult

This practical can be used as a demonstration, as the energy transfer between elastic potential energy stores and kinetic energy stores should be explained separately and carefully: these are two concepts, and linking these stores requires two equations. There is a risk of cognitive overload, so scaffold delivery by focusing initially on the energy transfer itself, without equations. Expect students to grasp that stretching the elastic in the catapult further stores more energy, which is transferred to the kinetic energy store of the projectile.

Review and practice

Kinetic cards revisited (10 mins)

The students should now be able to calculate the exact kinetic energy of each of the cards and confirm their original ordering. They can use this task to self-assess their treatment and understanding of the equation.

Extend this activity by exploring ideas about kinetic and elastic potential energy stores as applied to the field of biomechanics, e.g., sports scientists studying the Achilles tendon, or in the design of effective shoes for professional athletes. Show students images and ask for their ideas about how a good understanding of these energy stores and transfers could be beneficial.

P1.6 Energy dissipation

AQA GCSE Physics specification points

4.1.2.1 Energy transfers in a system

Smart Curriculum Narrative

Energy

Theme/subtheme

Energy/Energy transfer

Disciplinary knowledge

- WS 1.4
- MS 3a

Practical

Investigating friction

	Learning outcomes		
Learning objectives	**Developing Aiming for 3**	**Secure Aiming for 5**	**Extending Aiming for 8**
Analyse energy changes to identify useful and less useful energy transfers.	Identify useful and wasted energy in simple scenarios	**Analyse energy changes to identify useful and less useful energy transfers.**	Use a wide range of energy stores and physical processes to decide on wasted and useful energy transfers
Describe energy dissipation and how this reduces the capacity of a system to do work	Describe energy dissipation in terms of heating the surroundings	**Describe energy dissipation and how this reduces the capacity of a system to do work**	Apply the concept of energy dissipation in a wide range of scenarios
Investigate the factors that affect frictional forces commenting on the precision of the results	Measure the frictional force acting on an object	**Investigate the factors that affect frictional forces commenting on the precision of the results**	Evaluate in detail an experiment to measure the frictional forces acting on an object

Tier 2 vocabulary	Tier 3 vocabulary
conserved, energy, force, friction, kinetic energy, Sankey diagram, *store*, system, transfer, transferred, wasted	dissipation, law of conservation of energy, thermal store

Lesson plan

Reactivate knowledge

1 Name a useful energy transfer of a phone.
2 Name an energy transfer of a phone which is not useful.
3 State what is meant by dissipated energy.

Common misconception

The phrase 'energy is lost' can lead to misconceptions that violate the law of conservation of energy. Remind students that energy is always conserved, so 'wasted' is a better way to describe non-useful energy that is transferred away from a system. This lesson is a great opportunity to emphasise the use of 'dissipation' as key terminology. Create a narrative through the lesson that energy is stored, usefully transferred between stores, or dissipated. Energy dissipation should be linked to the idea of it 'spreading out', wherever possible. Dissipation does not have to relate to thermal energy stores – it simply means the energy pathway during transfer widens to other unwanted energy stores.

Answers: 1 Light/sound energy *2* Energy transferred to thermal store *3* Energy transferred to less useful stores

Trigger interest

Braking in the W Series (10 mins)

Show a video clip of W Series highlights, asking students to discuss the energy transfers involved in the cars as they brake into corners. Links can be made about friction between the tyres and the track, as well as between the brakes and the tyres. Lead students to conclude that the surroundings are heated, and that the energy in the resulting thermal stores of the brakes, the tracks, or the air cannot be usefully reused.

Exposition

Energy pathways (10 mins)

Discuss some example energy transfers, visualised differently, for example, some as Sankey diagrams, some as images with key energy stores described. Quantitative analysis is not important here – students can use cards with corresponding letters to vote for useful transfer pathways (U) and wasteful transfer pathways (W).

Main activity

Investigating friction (25 mins)

Students can observe or attempt the practical, noting that heating of the surroundings is the ultimate effect of most energy transfers, and that force is the mechanism by which energy is transferred.

Review and practice

Spot the misconception (10 mins)

Ask the students to correct some sentences describing energy and friction. This can be used to assess their understanding, praise use of key terms, and address continued misconceptions. Examples can include: 'When a car stops at traffic lights, the [kinetic energy] [is destroyed] by the brakes and is [lost].'

P1.7 Energy and efficiency

AQA GCSE Physics specification points

4.1.2.2 Efficiency

Smart Curriculum Narrative

Energy

Theme/subtheme

Energy/Energy transfer

Disciplinary knowledge

- WS 1.4 (H), 4.3, 4.4, 4.5, 4.6
- MS 1c, 3a, 3b, 3c

Practical

Investigating efficiency

	Learning outcomes		
Learning objectives	**Developing Aiming for 3**	**Secure Aiming for 5**	**Extending Aiming for 8**
Calculate the efficiency of a range of energy transfers	Calculate the efficiency of a simple energy transfer	**Calculate the efficiency of a range of energy transfers**	Rearrange the efficiency equation to find input or total output energy
Explain ways of reducing unwanted energy transfers	Describe an efficient transfer as one that transfers more energy usefully	**Explain ways of reducing unwanted energy transfers**	(H) Describe design features that can be used to improve the efficiency of an energy transfer
Identify measurements needed to find the efficiency of a process	Determine the efficiency of a process from measurements	**Identify measurements needed to find the efficiency of a process**	Evaluate the method to improve reliability and accuracy of measurements

Tier 2 vocabulary	Tier 3 vocabulary
calculation, emission, energy, equation, input, metric, motor, output, percentage, Sankey diagram, *store*, transfer, useful, *waste*	efficient, efficiency, work done

Lesson plan

Reactivate knowledge

1 State the useful energy transfer of a light bulb.

2 State an unwanted energy transfer of a light bulb.

3 How does the energy input to a light bulb compare to the total energy output?

Common misconception

The phrase 'energy is lost' can lead to misconceptions that violate the law of conservation of energy. Remind students that energy is always conserved, so 'wasted' is a better way to describe non-useful energy that is transferred away from a system. This will appear explicitly in applications of the efficiency equation.

Answers: 1 Light *2* Heating/energy transfer to thermal store *3* They are the same

Trigger interest

Car efficiency (5 mins)

Introduce the idea of efficiency using the metric of 'miles per gallon'. Ask students if they are familiar with this term and set up the central idea that more miles per gallon (mpg) is better, because it indicates a better efficiency of energy transfer.

Students can carry out research to compare two cars of differing efficiencies (ideally using mpg) and discuss their prices in the context of other things, like CO_2 emissions ratings. You can extend this by considering other units, for example, miles per kWh.

Exposition

Calculating efficiency (15 mins)

Sankey diagrams are useful as visual tools. To set up the idea of efficient energy transfers, compare two generic diagrams, only labelling the starting store A, and two final stores B, and C.

In abstract terms, describe a machine or process that aims to transfer energy from A to B. Students should quickly understand that the diagram in which more energy ends up in store B represents the more efficient machine/process. Build on this abstraction by labelling starting store A as the total input and store B as the useful output, with students concluding that store C is called the waste. Annotate this further as appropriate and necessary (e.g., specific energy stores), emphasising the processes involved in the energy transfers.

Introduce the equation, following which students should perform a few calculations to ensure they can perform them. Extend this by challenging students to express their answers as a percentage.

Main activity

Investigating efficiency (30 mins)

Students try an efficiency measurement using the practical activity. Ensure that they are calculating work done correctly, and finding the energy supplied to the motor. This is a good opportunity to discuss the nature of the waste here via heating in the motor.

Review and practice

Improving car efficiency (5 mins)

Return to some of the cars found at the beginning of the lesson. Brainstorm, as a whole-class discussion, the design features used to improve efficiency of cars. Expand this discussion to include wider issues around clean energy, and the global push toward electric vehicles.

P1.8 Electrical appliances

AQA GCSE Physics specification points

4.1.1.1 Energy stores and systems

4.1.2.2 Efficiency

Smart Curriculum Narrative

Energy

Theme/subtheme

Energy/Energy stores

Disciplinary knowledge

- WS 1.2, 1.4
- MS 1a, 1c, 3a, 3b, 3c

Activity

Mains- or battery-powered

	Learning outcomes		
Learning objectives	**Developing Aiming for 3**	**Secure Aiming for 5**	**Extending Aiming for 8**
Describe the energy transfers carried out by electrical devices	List some example electrical devices	**Describe the energy transfers carried out by electrical devices**	Explain the advantages and disadvantages of electrical devices
Describe why electrical appliances are useful	Survey a range of electrical devices and their operation	**Describe why electrical appliances are useful**	Compare mains-powered and battery-powered devices
Describe the processes that waste energy in electrical devices	Identify waste energy transfers in electrical devices	**Describe the processes that waste energy in electrical devices**	Compare electrical devices in terms of efficiency

Tier 2 vocabulary	Tier 3 vocabulary
analyse, appliance, efficient, electrical, electricity, energy, temperature, transfer	battery-powered, efficiency, electrical current, electrical resistance, mains-powered

Lesson plan

Reactivate knowledge

1 Name two ways of heating food.

2 Name an electrical appliance that transfers energy to a kinetic store.

3 Name an electrical appliance where heat is transferred usefully *and* as an unwanted energy transfer.

Common misconception

Students may think that a blown fuse explodes, or 'blows up', inside electrical appliances, if the electrical current is too large. Tell students that a fuse is, at its simplest, just a wire. All wires get warm when a current flows through them, due to electrical resistance: if this current gets too large, the resistance is so high that the high temperature causes the wire to melt. This is how a fuse breaks a circuit and keeps an appliance safe from damage.

Answers: 1 By electricity, by burning a fuel such as wood/gas/oil *2* Any appliance with a motor *3* Any heating appliance, e.g., kettle, iron

Trigger interest

Electricity everywhere (10 mins)

Ask students, in groups, to list all of the electrical appliances that they use every day. They will only score a point if their group can suggest something that no other group gets – this will encourage lateral thinking. Extend this activity by asking groups to explain the use of each appliance, and then to categorise their suggestions in terms of both mains-powered and battery-powered devices.

Exposition

Demonstration: Everyday appliances (10 mins)

Carry out the practical demonstration to show the possible effects of an electric current. In the first demonstration, showing a glowing resistance wire, make sure that students link this to how fuses work if a current gets too large.

Main activity

Mains- or battery-powered (25 mins)

Provide students with images of electrical appliances in different rooms in a typical home. They could use coloured pens or patterns to identify and label mains-powered and battery-powered devices, leading to considerations about rate of energy transfer (power).

Make students aware of clockwork devices and show them one, if possible.

Students then analyse the different devices mentioned in the Student Book, discussing how they operate, and link this back to the effects of a current from the demonstration earlier in the lesson.

Review and practice

Making connections (10 mins)

Students complete the paragraph:

Electrical current is a very convenient way of transferring energy because....

They should include the words energy, transfer, and current. Students then calculate the efficiency of a series of electrical appliances, ranking them from the most to the least efficient.

P1.9 Energy and power

AQA GCSE Physics specification points

4.1.1.4 Power

4.1.2.2 Efficiency

Smart Curriculum Narrative

Energy

Theme/subtheme

Energy/Energy efficiency

Disciplinary knowledge

- WS 1.2, 4.3, 4.4, 4.6
- MS 1a, 1b, 1c, 3a, 3b, 3c

Practical

Efficiency and power

	Learning outcomes		
Learning objectives	**Developing Aiming for 3**	**Secure Aiming for 5**	**Extending Aiming for 8**
Calculate the power of an appliance	Define power and give its unit	**Calculate the power of an appliance**	Rearrange the power equation to determine the energy transferred or the work done
Calculate the efficiency of an appliance from power ratings	Identify the power input and useful power output for an appliance	**Calculate the efficiency of an appliance from power ratings**	Apply the efficiency equation in a range of situations, including rearrangement of the equation
Identify the more powerful device from data	With support, rank electrical appliances in order of power	**Identify the more powerful device from data**	Compare the power ratings of devices using standard form

Tier 2 vocabulary	Tier 3 vocabulary
appliance, calculation, efficiency, electrical, energy, evolutionary, joule, meter, power, *rate*, seconds, symbol, transfer, wasted	joulemeter, power output, power rating, watt (W), work done

Lesson plan

Reactivate knowledge

1 Describe the difference between energy and power.

2 How do you calculate power?

3 State how you calculate the efficiency of a device.

Common misconception

Students may think all units displayed on electrical appliances are various ways of showing 'power consumption'. Since power × time = energy, then this is a unit of energy. It is typically used in homes to describe the amount of energy a person would use if they kept a 1000 W appliance running for one hour.

Answers: 1 Power is the rate of transfer of energy *2* Power = $\frac{\text{energy transferred}}{\text{time taken}}$ *3* Efficiency = $\frac{\text{useful output energy transfer}}{\text{total input energy transfer}}$

Trigger interest

Match up (10 mins)

Create a card sort activity that matches a range of electrical appliances to their power rating. If required, display a short definition of power to guide students' intuition, e.g., the power of an appliance is how much energy it uses every second. Since this is the start of the lesson, reassure students that it is good to discuss and use their intuition to make some educated guesses, and that they will be able to revisit their suggestions later.

Extend this by asking students to identify the useful energy transfers and wasted energy transfers of each appliance.

Exposition

Efficiency and power calculations (20 mins)

Introduce the power equation in words, and then in symbols. Remind students that 'energy transferred' is equivalent to saying 'work done'. Immediately revisit some of the appliances from the Match up activity as examples, translating their power rating into words, for example, 60 W means the appliance transfers 60 joules per second. Next, go through a few of the examples from the Student Book, so that they see each stage of a typical calculation. They should they try an example of their own. A maths skills interactive is available to support students with the calculations. Also provide some examples for them to carry out themselves.

Students then identify the useful and wasted power output of a range of devices and then use this data to find the efficiency. Emphasise careful layout of calculations, paying particular attention to unit prefixes, to avoid mistakes.

A support sheet is available where students develop their knowledge of the units of energy and power and also to practise using energy terms.

Main activity

Efficiency and power (15 mins)

Ahead of this demonstration, introduce students to the function of a joulemeter, leading them to appreciate that it measures the transfer of energy with emphasis on 'joule' and 'meter'.

Carry out the demonstration to compare the difference in energy consumption of two electrical appliances in 60 seconds. Ideally, these will be two of the appliances investigated from the previous lesson or starter activity, for example, a kettle and a motor. Students should quickly appreciate that one appliance has a higher power, or rate of energy transfer, than the other. Gently allude to the idea that it is useful to compare appliances in terms of how much energy they transfer in the same unit of time, which is typically in one second.

Review and practice

Low-power mode (10 mins)

Lead a whole-class discussion about battery-life in modern smartphones, and the proliferation of the low-power mode, or power-saving mode, to conserve battery life. Expect students to be familiar with the effects of power-saving, for example, a dimmer screen, no push-notifications, and so on. They should be able to describe these power-saving mechanisms in terms of energy transfers and stores to different functions of the phone.

Extend this discussion to encourage students to explore whether or not the animal brain has an equivalent low-power mode to that of modern smartphones, and discuss the benefits of this characteristic, with respect to the evolutionary responses, to the risk of having to go for long periods/seasonal periods without food.

P1 Conservation and dissipation of energy

Introduction to checkpoint intervention

The checkpoint intervention provides suggestions for a lesson to follow up the *P1 Conservation and dissipation of energy* checkpoint assessment. Depending on the outcomes of the assessment, these suggestions can be used to move a class towards secure, to target specific learning outcomes, or to consolidate knowledge for students achieving secure outcomes. Use students' outcomes from the checkpoint assessment to address general misconceptions from the content covered. Be prepared to re-cover content as required; three key concepts from the chapter are covered in more detail below.

Checkpoint secure learning outcomes

- Use the quantitative descriptions of changes in energy stores.
- Describe the energy changes in a range of experiments and account for energy dissipation to the surroundings.
- Use the equation for work done to calculate distances or sizes of forces.
- Calculate the kinetic energy store of an object.
- Calculate the elastic potential energy store of a stretched spring.
- Analyse energy changes to identify useful and less useful energy transfers.

Student reflection

Encourage students to reflect on whether there were any checkpoint questions they found difficult or straightforward as well as their preparation for the checkpoint assessment, including how they revised, what they revised, the time spent on revision, and what areas they could improve upon. For more strategies, see the Metacognition spread.

Intervention activity: Use the quantitative descriptions of changes in energy stores.

The idea of lost energy is a stubborn misconception that students should be quick to identify and correct. Emphasise the idea that when energy spreads out, it normally means it goes down unwanted pathways. This can represented using a simple analogy that gets students to select pathways based on desired endpoints. For example, sketch a car on a main road that splits into three different destinations, e.g., a petrol station, a restaurant, and a bowling alley. If the person driving would like to buy dinner, ask students which pathway and endpoint is likely to be preferred by the driver, and get them to highlight said route from start to finish. In a similar way, in the context of energy transfers, students should then be able to highlight useful energy pathways by recognising these as 'preferred' pathways, e.g., from chemical stores to kinetic stores in a jet, because jets want to move. Then, move on to highlighting non-useful, or wasted pathways (e.g., from chemical stores to thermal stores as the jet moves along the runway).

Students can label or use coloured pens to show the energy pathways for energy transfers, between the starting store and the final stores in which the energy ends up. This can lead nicely on to ideas about how to reduce unwanted dissipation, for example, through insulation, and into broader ideas about efficiency.

Intervention activity: Describe the energy changes in a range of experiments and account for energy dissipation to the surroundings.

The idea that energy appears in different contexts, e.g., hot and cold objects, moving objects, etc, is usually understood by students, but framing this in language about energy stores can be more challenging. The simplest scenario of a falling object, demonstrating the transfer of energy from its gravitational potential energy store to its kinetic energy store, is useful to practise this language.

A falling ball is a useful and simple demonstration to scaffold this concept. Draw buckets to represent the gravitational potential energy store and kinetic energy store of the ball at different points on its descent. Emphasise that the buckets themselves are not real – they help us imagine the energy stores of the object; thus, students should be able to illustrate by colouring/shading the buckets filling/emptying at the top, middle, and end of the ball's descent.

The aim is to develop this idea by introducing the notion that the ball is in a closed system, ignoring everything else (i.e., the air and the ground). If these things are included in the system under consideration, then students should be able to draw and fill buckets for the thermal stores of the air and the ground themselves.

Intervention activity: Analyse energy changes to identify useful and less useful energy transfers.

Some students might not immediately grasp the link between the efficiency equation based on energy transfers, and the efficiency equation based on power rating.

It should be emphasised that both equations calculate the same thing, but the information we are given for electrical appliances is not merely how much energy they transfer, but how quickly they transfer that energy. It is useful to split students into two groups to compare the value they calculate for efficiency using energy transfers, and for power, expecting them to get the same answer. Keep the scenario simple. For example, a torch lamp, with 100 J in the chemical energy store in a battery, and with transfers of 20 J towards light waves and 80 J towards the thermal energy store of the surroundings. Lead students to see that dividing these energy transfers by the same unit of time, will leave the final value for efficiency unchanged.

P2 Energy transfer by heating

P2.1 Energy transfer by conduction

AQA GCSE Physics specification points
4.1.2.1 Energy transfers in a system

Smart Curriculum Narrative
Energy

Theme/subtheme
Energy/Energy transfer

Disciplinary knowledge
- MS 2c, 4a

Required Practical 2
Thermal insulation

	Learning outcomes		
Learning objectives	**Developing** **Aiming for 3**	**Secure** **Aiming for 5**	**Extending** **Aiming for 8**
Analyse data to compare the thermal conductivity of materials	Describe materials as good thermal conductors or insulators	**Analyse data to compare the thermal conductivity of materials**	Explain the different thermal conductivities of materials using the free electron and lattice vibration explanations of conduction
Describe how the rate of cooling of a building depends on thermal conductivity and thickness of the materials used	State the factors affecting the rate of cooling of a building	**Describe how the rate of cooling of a building depends on thermal conductivity and thickness of the materials used**	Explain how a building can be designed to maintain its temperature with little heating or cooling
Design a fair test to investigate the effectiveness of different insulators	Identify independent, dependent, and control variables	**Design a fair test to investigate the effectiveness of different insulators**	Evaluate the results of a test to investigate the effectiveness of different insulators in terms of the validity of conclusions drawn from the data

Tier 2 vocabulary	Tier 3 vocabulary
Bunsen burner, conduction, *conductor, degree* Celsius, energy, fair test, insulate, metal, non-metal, temperature, thermometer	energy transfer, thermal conductivity, thermal insulator

Lesson plan

Reactivate knowledge

1 Describe the arrangement of particles in a solid.

2 Describe how the particles in a solid behave when the solid is heated.

3 What is conduction?

Common misconception

Students may think that insulators do not conduct energy. Although it is useful to group materials into the category of conductor or insulator, in reality, all materials fall on a spectrum of conductivity. For instance, wood does get warm, just not as quickly as copper.

Answers: 1 Particles are fixed in place and vibrate *2* Particles vibrate more *3* The transfer of energy by vibrating particles colliding with other particles

Trigger interest

To fry, or not to fry? (10 mins)

Set up two Bunsen burners behind a safety screen – one under a frying pan and one under a heat resistant mat. With equal flames, judged by eye, heat the frying pan and the heat resistant mat, asking students to time 60 seconds to ensure a fair test (i.e., the same amount of energy transferred to each material). Ask the class to predict which surface will fry an egg, justifying their predictions by describing the processes that allow energy to reach the eggs and the rate of energy transfer.

Fry an egg on each surface to check their predictions.

Extend the discussion by asking students to imagine zooming in on the particles that make up the frying pan and the heat-resistant mat – describing the difference in their motion as they are heated.

Exposition

Melting wax (10 mins)

Use Figure 1 in the Student Book to set up a thought experiment. Ask students to vote whether the material with the wax end that melts the fastest is the best or worst conductor. To extend students, ask them to suggest material types for each rod.

Emphasise the need for adequate controls for a fair test (i.e., the amount of wax, the length of the rod, and the proximity to the flame).

Use this thought experiment to introduce the key concept of thermal conductivity and insulators. Introduce examples (images are helpful). For example, contrast the similarities between loft insulation and large 'puffer' jackets.

Main activity

Thermal insulation (30 mins)

Students complete the required practical, focusing on producing results that are as accurate as possible by refining the control variables and eliminating sources of random error.

Links should be made to the idea that trapped air in the foams or fluffy materials used for insulation reduces energy dissipation. This can be visualised as a Sankey diagram by a narrowing energy pathway to thermal stores of the surroundings.

Extend students by asking them to explore the quantitative relationship between the number of layers of insulation with temperature change. They should, in this case, identify the independent variable (number of layers), dependent variable (temperature change), and their control variables (e.g., container, amount of water).

Review and practice

Are you a robot? (5 mins)

Create a Google-style bot-check grid of nine images. Each image should show objects made of different materials (e.g., metal, glass, or plastic). Students, in pairs, should select the three images that show materials with the highest thermal conductivity.

Ask them to check their answers with you to confirm they are human and not robots!

Support/extend this activity by carefully selecting images, including obvious and ambiguous materials as appropriate.

P2.2 Specific heat capacity

AQA GCSE Physics specification points

4.3.2.2 Temperature changes in a system and specific heat capacity

Smart Curriculum Narrative

Energy

Theme/subtheme

Energy/Energy and temperature

Disciplinary knowledge

- MS 1a, 3b, 3c, 3d

Required Practical 1

Specific heat capacity

	Learning outcomes		
Learning objectives	Developing Aiming for 3	Secure Aiming for 5	Extending Aiming for 8
Describe what specific heat capacity is	List the factors that affect the amount of energy required to increase the temperature of an object	**Describe what specific heat capacity is**	Evaluate materials used for transferring energy in terms of their specific heat capacity
Calculate the change in thermal energy when there is a temperature schange	State that the change in thermal energy stored is equal to the energy transferred	**Calculate the change in thermal energy when there is a temperature change**	Use the specific heat capacity equation to perform a wide range of calculations in unfamiliar contexts
Measure the specific heat capacity of a material and find a mean value	With some support, measure the specific heat capacity of a material	**Measure the specific heat capacity of a material and find a mean value**	Evaluate in results of an experiment to measure specific heat capacity, including comparing to an actual value

Tier 2 vocabulary	Tier 3 vocabulary
energy, joule, joulemeter, kilogram, mass, temperature	energy transfer, specific heat capacity, thermal energy store

Lesson plan

Reactivate knowledge

1 Describe the intended changes to energy stores when a kettle is switched on.

2 Name two things the temperature rise of a substance depend on.

3 State the equation relating energy transferred, power, and time.

Common misconception

Students may think that 'heat' is the same as 'temperature'. Remind students that temperature describes how hot an object is. Heating is the process by which energy is transferred between the thermal stores of two objects at different temperatures.

Answers: 1 Energy transferred by electric current to heater, heater transfers energy by heating to water, and thermal store of water increases *2* Energy transferred, mass of the substance, and what the substance is *3* Energy transferred = power × time

Trigger interest

Boiling challenge (10 mins)

Introduce a thought experiment to compare two identical kettles, A and B, in a race to make the fastest cup of tea. Kettle A has exactly the amount of water required for a single cup, whereas B is filled with far more water. Students should predict intuitively that kettle A wins. Explore students' rationale for this prediction, leading them to conclude that the amount of a substance is important when deciding how much energy is required to raise its temperature.

Extend this scenario by introducing a third identical kettle (C) filled with the same amount of cooking oil as water in kettle A (with a do-not-try-at-home warning!). Now ask students to predict which substance reaches 100°C first. Students could discuss in pairs, and vote for kettle A, kettle C, or both at the same time.

Explore students' interests or experiences with cooking – boiling and/or frying – whilst introducing the notion that oil heats up much faster than water, and hence the type of substance is also important when deciding how much energy is required to raise its temperature. Reveal that the same amount of oil reaches 100°C in about half the time it takes water.

Show students an energy forms and changes PhET simulation similar to the previous scenario. There is opportunity here to link to broader ideas about responsible energy usage (e.g., only boiling the amount of water necessary for the purposes of drinking or cooking).

Exposition

Measuring specific heat capacity (10 mins)

Define the specific heat capacity, c, in words to help students understand the meaning of each symbol in the equation. Consider annotating each symbol with its definition.

Ensure students understand that specific heat capacity depends on the type of substance. You could link back to the thought experiment here, and contrast the specific heat capacity of water and oil. Remind students who need additional support about the meaning of symbols like ΔE and $\Delta\theta$.

Reinforce students' intuition before leading them through a calculation. Using the specific heat capacities in Table 1 of the Student Book, ask students to vote, and then explain, which substance gets hotter by 1°C quickest if supplied with 100 J of energy. An example, showing each of the steps using the equation in the form of a simple calculation frame, $\Delta E = mc\Delta\theta$, will help students in need of additional support. The extension sheet also provides further practice on rearranging the equation.

Main activity

Specific heat capacity (30 mins)

Students then carry out the required practical to find the value for the specific heat capacity of a metal.

Display the accepted value of the metal's specific heat capacity. Create some competition by asking students to predict if their measured value will be equal to, higher, or lower than this. Circulate around the groups as they measure the temperature change of the metal, scaffolding questions appropriately to allude to the thermal energy stores of both the metal and the surroundings.

Expect students' measured values to be greater than the accepted value; mainly due to energy being dissipated into the surroundings, as well as into the metal block itself.

Review and practice

Crossword (5 mins)

Students use the interactive to complete a crossword on the content from this lesson and previous lessons. This should form a useful summary of their learning about energy transfer.

P2.3 Heating and insulating buildings

AQA GCSE Physics specification points

4.1.2.1 Energy transfers in a system

Smart Curriculum Narrative

Energy

Theme/subtheme

Energy/Energy transfer

Activity

Reducing energy transfers in the home

	Learning outcomes		
Learning objectives	**Developing** **Aiming for 3**	**Secure** **Aiming for 5**	**Extending** **Aiming for 8**
Describe how some design features used to reduce energy dissipation from a home work	Describe some design features used to prevent energy transfer to the surroundings in the home	**Describe how some design features used to reduce energy dissipation from a home work**	Explain how design features used to reduce energy transfers from buildings work
Compare home improvement features in terms of payback time	Calculate the payback time of a simple home improvement feature	**Compare home improvement features in terms of payback time**	Decide on home improvement features using payback time and savings beyond the payback time

Tier 2 vocabulary	Tier 3 vocabulary
conduction, energy, fibreglass, insulation, insulator, mass, radiation, reflect, solar panel, Sun, *volume*	convection, density, energy transfer, heat pump, infrared radiation, specific heat capacity, thermal conductivity

Lesson plan

Reactivate knowledge

1 Do insulating materials have a high or low thermal conductivity?

2 How does the thickness of a wall affect the rate of energy transfer?

3 How can wrapping insulation around a can of ice cold soda help keep it cool?

Common misconception

Students may think that insulating buildings is only useful in winter. Emphasise that insulation not only keeps buildings warm during the winter, it also keeps them cool during the summer. This is because in both cases, insulation restricts the transfer of energy: in winter, it restricts energy transfer to thermal stores outside, whereas in summer, it restricts the transfer to thermal stores inside.

Answers: 1 Low thermal conductivity *2* The thicker the wall the lower the rate of energy transfer *3* It reduces the rate of energy transfer from outside to inside the can

Trigger interest

Energy and homes (5 mins)

Ask students to raise their hands if they have heard of or seen an eco-house. Students may be able to define one as an environmentally low-impact home designed and built with materials and technology that reduces its carbon footprint and has low energy needs.

From this definition, various conversations and potential misconceptions could be explored, e.g., What types of materials and/or technology are eco-friendly? What is carbon footprint? Why is low-energy needs a good thing for the person living in the home and for society?

Explore some features of an example eco-house from an internet search and explore some things they may have in common, in terms of minimising dissipated ('waste') energy, e.g., smart technology, reclaimed wood, insulation, natural lighting.

Exposition

Hot house (20 mins)

Provide students with a set of images of large homes, flats, and other buildings. Students should discuss where they predict the largest energy transfers to the surroundings will occur, annotate these on the images directly, and suggest the ways in which this energy dissipation can be minimised.

Next, provide images of internal and external infrared images of houses and other buildings. Students should use these images to evaluate their predictions, identifying where the energy transfer 'hotspots' are, as revealed by infrared viewing, for example, around the edges of windows.

Main activity

Reducing energy transfers in the home (25 mins)

Discuss the features used to prevent energy transferred in a typical home, one by one.

If available, pass around some sample materials (e.g., a brick, some insulating foam). Provide students with some examples of costs, as these will be useful for the House analysis activity. Students can discuss the relative costs of typical household methods to reduce energy transfers, e.g., double glazing, secondary glazing (glass or film), or thick curtains for a window. Be aware that students may have very different perceptions about what a 'typical' home looks like, so inclusive language and examples should be considered wherever possible, e.g., using local homes as examples.

Review and practice

House analysis (5 mins)

Provide a set of home improvements costs and savings. Students use the interactive to sequence them in the order in which improvements should be done.

P2 Energy transfer by heating

Introduction to checkpoint intervention

The checkpoint intervention provides suggestions for a lesson to follow up the *P2 Energy transfer by heating* checkpoint assessment. Depending on the outcomes of the assessment, these suggestions can be used to move a class towards secure, to target specific learning outcomes, or to consolidate knowledge for students achieving secure outcomes. Use students' outcomes from the checkpoint assessment to address general misconceptions from the content covered. Be prepared to re-cover content as required; three key concepts from the chapter are covered in more detail below.

Checkpoint secure learning outcomes

- Design a fair test to investigate the effectiveness of different insulators..
- Outline the evidence that changes in the concentration of atmospheric gases are the likely cause of global warming.
- Describe what specific heat capacity is.
- Measure the specific heat capacity of a material and find a mean value.
- Compare home improvement features.

Student Reflection

Encourage students to reflect on whether there were any checkpoint questions they found difficult or straightforward as well as their preparation for the checkpoint assessment, including how they revised, what they revised, the time spent on revision, and what areas they could improve upon. For more strategies, see the Metacognition spread.

Intervention activity: Design a fair test to investigate the effectiveness of different insulators.

The role of insulating material is intuitive to students, but descriptions about how they minimise energy dissipation is more challenging. This intervention activity focuses on use of key terminology.

Allow students to choose an insulating material from a selection of everyday items; for example, a piece of foam, plastic, or a fibrous blanket. They should complete a gap-fill paragraph to describe how their material acts as a thermal insulator, ideally as a four- or six-mark question, using key terms provided (e.g., thermal conductivity, insulator, dissipation). Ensure answers are peer-assessed to reinforce their understanding.

Intervention activity: Outline the evidence that changes in the concentration of atmospheric gases are the likely cause of global warming.

You can expect students to be familiar with the idea that the Earth's temperature is rising. This intervention activity can be conversational, as well as instructional – explore students' awareness of issues and opinions on the effects of global warming (and why they should care).

Create a set of cards, each having a greenhouse gas labelled (e.g., carbon dioxide, methane, nitrous oxide, and CFCs). Each card could have an image of each gas (ideally, where they come from). Emphasise that what distinguishes greenhouse gases from other atmospheric gases is their ability to absorb infrared radiation emitted from the Earth, keeping it warm. Students should then select one of the greenhouse gases and carry out independent research into one way that humans have contributed to its increase in the atmosphere in modern times. Their conclusion should be that a higher average global temperature is likely to be the result of human activity.

Intervention activity: Describe what specific heat capacity is.

The role of insulating material is intuitive to students, but descriptions about how they minimise energy dissipation is more challenging. This intervention activity focuses on use of key terminology.

Compare two exaggerated scenarios for each of the components involved in the equation $\Delta E = mc\Delta\theta$ to demonstrate their proportionality to required energy ΔE. Make these as relevant to everyday life as possible, and use visual prompts. For example, ask students to vote for the scenario that requires more energy – heating an outdoor lido by 1°C or by 100°C, reheating one slice of pizza or a whole pizza, heating a metal spoon or a wooden spoon? Establish that the energy required ΔE is proportional to m, c, and $\Delta\theta$. Provide plenty of example calculations to practise substitution and emphasise the units of each quantity. Example calculations, e.g., for water, the specific heat capacity is 4200 J/kg/C, and for lead it is much less, at 129 J/kg/C.

P3 Energy resources

P3.1 Energy demands

AQA GCSE Physics specification points

4.1.3 National and global energy resources

Smart Curriculum Narrative

Energy

Theme/subtheme

Energy/Energy resources

Disciplinary knowledge

- WS 4.4
- MS 1b

Activity

Using fuels for energy

	Learning outcomes		
Learning objectives	**Developing Aiming for 3**	**Secure Aiming for 5**	**Extending Aiming for 8**
Describe how different fuels are used for transport, heating and generating electricity	State the main uses of energy resources	**Describe how different fuels are used for transport, heating, and generating electricity**	Compare energy use from different sources and different societies from available data
Describe how power stations use fuels to generate electricity	Identify which fuels are renewable and which are non-renewable	**Describe how power stations use fuels to generate electricity**	Compare fossil fuels and nuclear fuels in terms of energy provided, waste, and pollution
Explain patterns in our use of energy resources	State the main energy resources used today	**Explain patterns in our use of energy resources**	Interpret data and explain patterns and trends shown

Tier 2 vocabulary	Tier 3 vocabulary
electricity, energy, gas, global warming, greenhouse gases, kilogram, *waste*	biofuel, coal, ethanol, fossil fuels, natural gas, nuclear fusion, nuclear power, oil

Lesson plan

Reactivate knowledge

1. Name two fossil fuels.
2. Define non-renewable energy resource.
3. Name a renewable fuel.

Common misconception

Students may think that energy is conserved so society cannot be 'running out of energy'. Emphasise that fossil fuels, which release energy when they are burned, are 'non-renewable', which means they cannot be replenished once used. Remind students that there are no energy 'types': energy is energy. The correct concept is that energy is transferred between different stores.

Answers: 1 Any two from coal, oil, or gas *2* Cannot easily be replaced *3* Any biofuel

Trigger interest

Fossil fuels (10 mins)

Use a set of images or a short video about the energy demands of a modern society, setting the scene with relatable ideas around transport, schools, offices, and homes (electricity for wi-fi, appliances, etc.). Students use the interactive to order the statements that describe how coal, oil, and natural gas are formed – ensure that students make the connection between the abstract (e.g., coal, oil, and natural gas) supply of the familiar energy source (electricity, gas).

Exposition

Power stations (15 mins)

There are three stations that you can walk students through, or allow them to explore independently in groups, making notes or answering prepared questions to scaffold their learning. Use videos or interactive animations, wherever possible, to demonstrate the operation of a (fossil fuel) power station, emphasising that most begin with the burning of fuel to heat water, creating steam. Contrast this with a short demonstration of the combustion of a biofuel (e.g., ethanol) while discussing how it is produced. Finally, nuclear power is best presented or shown through video – there is great opportunity here to extend students' curiosity by linking to recent advances in nuclear fusion.

Emphasise the common end-goal – that all of these processes eventually go on to generate electricity (it's ok at this stage to allude to the fact that something needs to 'turn' in order to generate electricity.)

Main activity

Using fuels for energy (15 mins)

Students complete the worksheet either by independent research, or after being presented with the information as above.

Review and practice

Advantages and disadvantages (15 mins)

Split the class into three groups, representing an oil-fired power station, a nuclear power station, and a coal-fired (fossil fuel) power station. Each group should try to convince the class that their method of energy production is best, using one or two advantages. Encourage students to think about metrics such as waste implications, energy released per kilogram of fuel, greenhouse gases (linking to prior learning about global warming). Extend the debate, if appropriate, to think about wider social/economic issues associated with each of method of energy production. This is a good opportunity to explore students' knowledge about them and share some differing perspectives.

P3.2 Energy from wind and water

AQA GCSE Physics specification points

4.1.3 National and global energy resources

Smart Curriculum Narrative

Energy

Theme/subtheme

Energy/Energy resources

Disciplinary knowledge

- WS 1.3, 1.4, 4.4
- MS 1c, 2c, 4a

Practical

Wind and water power

	Learning outcomes		
Learning objectives	**Developing Aiming for 3**	**Secure Aiming for 5**	**Extending Aiming for 8**
Describe how water and wind can be a source of energy	State that wind turbines, wave generators, hydroelectric systems, and tidal systems are renewable energy resources	**Describe how water and wind can be a source of energy**	Compare the operation of hydroelectric, wave, and tidal systems in terms of reliability, potential power output, and costs
Describe advantages and disadvantages of hydroelectricity, wind, wave, and tidal power	Describe some simple advantages or disadvantages of renewable energy systems	**Describe advantages and disadvantages of hydroelectricity, wind, wave, and tidal power**	Present arguments for and against hydroelectric, wave, and tidal systems from different perspectives
Suggest the most appropriate energy resource to use in a range of scenarios	State the use of wind and water as a resource for transport and generating electricity	**Suggest the most appropriate energy resource to use in a range of scenarios**	Justify the choice of an energy resource for a particular use by using numerical and other appropriate data

Tier 2 vocabulary	Tier 3 vocabulary
atmosphere, climate, convection, efficiency, electric current, energy, fossil fuel, *power, pressure*	gravitational potential, kinetic, power station, wind farm, wind turbine

Lesson plan

Reactivate knowledge

1. Define renewable energy resource.
2. Name an energy resource that uses water.
3. Give one reason we cannot rely completely on fossil fuels.

Common misconception

Students may think that wind and tidal power have no environmental disadvantages as they are renewable sources of energy. Tell students that some people may not like the look of large wind turbines and wind farms imposed on the natural landscape. Additionally, tidal power may not be produced reliably and at times when there is peak demand.

Answers: 1 Can be replaced *2* Hydroelectricity, tidal, or wave *3* Any from: Releases stored carbon dioxide into atmosphere, causing climate change

Trigger interest

World's wind (10 mins)

Display an image of the red spot on Jupiter to trigger interest about planets with atmospheres that experience wind (storms!) and convection currents. Contrast this with Earth, by providing students with a basic diagram showing an ocean, land, mountains, and a river. Ask students to discuss and annotate the diagram with ideas about how wind is generated on Earth; extra support can be provided by giving an order to the process, for example: (1) The Sun heats the Earth's surface, (2) Land heats up faster than water, and so on. Extend this conversation by going into more depth about causes of air flow in terms of expansion, density changes, and pressure – leading to a recognition of wind causing waves which then carry energy across the ocean.

Exposition

Wind and water (10 mins)

Show a video of a wind farm and ask students to explain what is happening. Link back to the idea of a turbine generating electricity. Discuss the advantages and disadvantages of using wind energy, including different locations of wind farms. Explore students' preconceptions about the prevalence of wind power; they could, using mini-whiteboards, estimate as a percentage how much of the UK's electricity was generated by wind (in 2023, this was 29.4%). There are lots of encouraging statistics to contextualise the transition to greater reliance on renewable energy sources like wind and water.

Next, show a video of a simple water turbine, pausing the video to discuss the energy transfers involved between kinetic energy stores and gravitational potential energy stores. To extend students, discuss the efficiency of pumped storage systems: emphasise that, although these systems are not efficient, they allow some of the excess power produced at night by the base load of the network to be used usefully.

Main activity

Wind and water power (25 mins)

Carry out the simple experiment to investigate wind turbines. Ensure that students understand that a voltage is equivalent to generating electricity, since a voltage drives the flow of electric current.

Ask students to predict the effect on the voltage output if the distance between the fan and the turbine increases, as well as if the number of turbine blades increases. Independent, dependent, and control variables should be identified.

Review and practice

Town planner (10 mins)

The students should choose either their hometown, or somewhere else they would like to live. They should describe the local environment, in terms of the typical climate, geographical location, and economic activity. Given this, they should suggest which of the energy generating systems covered in this lesson would best suit their town.

Extend their responses by asking if the systems are a better solution to local needs than a fossil fuel power station. Support students, where needed, by providing example towns and their local environment, with propositions in place, so they can evaluate these directly.

P3.3 Power from the Sun and the Earth

AQA GCSE Physics specification points

4.1.3 National and global energy resources

Smart Curriculum Narrative

Energy

Theme/subtheme

Energy/Energy resources

Disciplinary knowledge

- WS 4.4
- MS 1c, 2c, 4a

Practical

Using solar cells

	Learning outcomes		
Learning objectives	**Developing Aiming for 3**	**Secure Aiming for 5**	**Extending Aiming for 8**
Plan an investigation into how the area of solar cell exposed affects voltage output	Explore the operation of a solar cell	**Plan an investigation into how the area of solar cell exposed affects voltage output**	Plan in detail an investigation into the factors that affect the voltage output of a solar cell
Describe the advantages and disadvantages of using solar energy	Describe how energy from the Sun can be used as an energy resource	**Describe the advantages and disadvantages of using solar energy**	Explain the advantage of using solar panels for heating water even in colder climates
Describe how geothermal energy can be used as an energy resource and its advantages and disadvantages	State that geothermal energy uses the heat inside the Earth	**Describe how geothermal energy can be used as an energy resource and its advantages and disadvantages**	Consider the environmental impact of geothermal schemes

Tier 2 vocabulary	Tier 3 vocabulary
absorption, control variable, density, dependent variable, electricity, energy, energy transfer, generator, independent variable, infrared radiation, pressure, reflection, renewable, thermal store	geothermal store, infrared heating, solar cell, solar heating panel, wind turbine

Lesson plan

Reactivate knowledge

1 Name one way you can use the energy from the Sun.

2 What is the name given to the energy resource that uses the energy from inside the Earth?

3 Which is the only energy resource to produce electricity without a generator?

Answers: 1 One from; solar cells, solar heating panels *2* Geothermal *3* Solar cells

Common misconception

Students may think that the centre of the Earth is hot because of high pressure / density. Inform students that the Earth's core is heated by the energy transferred through radioactive decays.

Trigger interest

To the centre of the Earth (10 mins)

Students use the interactive to label a simple diagram showing the layers of the Earth (crust, mantle, and core), matching each layer with its properties. They should be aware that the outer core is a molten layer, and that the centre of the Earth is very hot – extend this by discussing comparisons to the centre of the Sun.

Exposition

Infrared heating (10 mins)

Remind students about ideas around heating (a process of energy transfer to thermal stores), temperature rise (an effect of that energy transfer), or by absorption of infrared radiation.

Discuss the renewable nature of geothermal systems – compare systems that use high pressure steam with ones that simply provide hot water for direct heating.

Emphasise that infrared radiation from the Sun is absorbed to create electricity directly (solar cells) or to heat water (solar heating panel), which indirectly creates electricity (high pressure steam to operate generators).

Main activity

Using solar cells (30 mins)

Students complete the practical to investigate how the voltage output of the motor will depend on the distance of the light bulb from the solar cell. Make time to allow students to plan their approach, linking to the similarities between this practical and the practical exploring wind turbines in the previous lesson. Support students by providing a simple framework to complete their plan, identifying the independent, dependent, and control variables.

Review and practice

Black and silver containers (5 mins)

This is a thought experiment, in which a lamp is placed near two equally-sized containers filled with water. One container is black, and the other is silver. Students should vote for which container will undergo the biggest temperature rise within a given time period. Encourage students to discuss their reasons with a partner, in terms of absorption and reflection of infrared radiation, before voting.

P3.4 Energy and the environment

AQA GCSE Physics specification points

4.1.3 National and global energy resources

Smart Curriculum Narrative

Energy

Theme/subtheme

Energy/Energy resources

Disciplinary knowledge

- WS 1.3, 1.4, 4.4
- MS 1c, 2c, 4a

Activity

Presenting issues

	Learning outcomes		
Learning objectives	**Developing Aiming for 3**	**Secure Aiming for 5**	**Extending Aiming for 8**
Describe the damage to the environment caused by burning fossil fuels	List some environmental problems associated with burning fossil fuels	**Describe the damage to the environment caused by burning fossil fuels**	Evaluate methods of reducing damage caused by waste products of fossil fuels
Describe the advantages and disadvantages of nuclear power	Identify the waste products of nuclear fuel	**Describe the advantages and disadvantages of nuclear power**	Discuss in detail the problems associated with nuclear accidents and the public perception of nuclear safety
Evaluate the use of different energy resources	Describe simple advantages and disadvantages of a variety of renewable energy resources	**Evaluate the use of different energy resources**	Evaluate the suitability of an energy resource for a range of scenarios, taking into account a wide range of factors

Tier 2 vocabulary	Tier 3 vocabulary
acid rain, decay, element, energy, environment, finite, renewable	fossil fuel, nuclear fuel, nuclear power, universal indicator, uranium

Lesson plan

Reactivate knowledge

1 Give one renewable and one non-renewable energy resource.

2 Name an energy resource which does not derive energy from the Sun.

3 Give an example of an unreliable energy resource.

Common misconception

Students may think that nuclear fuel is a renewable source of energy. Emphasise that nuclear fuel, like the element uranium, is not renewable because it is finite in supply and does not get replenished while it is being used (i.e., once it decays).

Answers: 1 Any fossil fuel or nuclear, and any suitable renewable *2* Any one from; nuclear, tidal, or geothermal *3* Any from; solar, wind, wave

Trigger interest

Demonstration: Acid rain (10 mins)

Use the practical to demonstrate the acidity of rainwater with a universal indicator. Ask students what may cause this, what the effects could be, and what the implications are for their environment.

There is opportunity in this lesson to connect the learning to students' lives and opinions, using more personal language (as in the Student Book), such as 'your' environment, rather than 'the' environment. This emphasises real-world relevance and encourages deeper engagement when reflecting on social and economic impacts of energy use and misuse.

Exposition

Researching fuel systems (15 mins)

Frame this as part one of a two-part series on the relationship between energy and the environment. Here, the aim is to gather information and fact-check, in order to form opinions to debate the broader issues.

Split students into three groups, tasked with research and presentation on fossil fuels, nuclear power, and renewable energy sources. Support groups as appropriate by providing a framework of questions that focuses their research – particularly emphasising the negative effects fossil fuels have on the environment.

Here, students are to develop research skills by gathering and quoting sources of information about the pros and cons of fossil fuels, nuclear power, and renewable energy sources. Encourage students to make propositions based on the evidence they find, and using their own lived experiences and opinions, using sentence frames like: "Based on [evidence] we think that [local people/businesses/industry/the government] should ..." Create your own as a model, which might also inspire some constructive debate amongst groups!

Main activity

Presenting issues (25 mins)

Have volunteers or small groups present their findings, followed by a panel-style question and answer session. Encourage students to ask panellists about both the advantages and disadvantages of each energy system.

Interject the discussion with a case study about the Chernobyl disaster and its effects, but also link to more recent accidents such as Fukushima Daiichi. This can be a brief video, or simply a series of images, emphasising the fact that accidents are very rare but can have significant and long-lasting damage.

Review and practice

What's the problem? (5 mins)

Remind students that they should have learnt facts about what fossil fuels do to their environment, the concerns about nuclear power and renewable sources, and how to evaluate these systems (by weighing up advantages and disadvantages). Use the interactive to match environmental problems with their likely causes.

P3.5 Big energy issues

AQA GCSE Physics specification points

4.1.3 National and global energy resources

Smart Curriculum Narrative

Energy

Theme/subtheme

Energy/Energy resources

Disciplinary knowledge

- WS 1.3, 1.4, 3.5, 4.4
- MS 1a, 1c, 2c, 4a

Activity

The big energy debate

	Learning outcomes		
Learning objectives	Developing Aiming for 3	Secure Aiming for 5	Extending Aiming for 8
Describe how different energy resources are used to meet variation in demand	State that demand for electricity changes during the day	**Describe how different energy resources are used to meet variation in demand**	Explain why some power stations are in constant operation whereas others may be switched on and off
Compare the costs of different energy resources	Describe how the cost of an energy resource depends on initial build costs, running costs, and final removal	**Compare the costs of different energy resources**	Use the costs of energy resources to evaluate their usefulness
Explain why the use of energy resources is changing	Describe how the use of energy resources is changing	**Explain why the use of energy resources is changing**	Discuss the difficulties of meeting global energy requirements in the future and meeting climate change targets

Tier 2 vocabulary	Tier 3 vocabulary
calculation, *demand*, electricity, energy, resource, supply	base-load, decommissioning, final cost, fossil fuel, initial cost, running cost

Lesson plan

Reactivate knowledge

1 Name a reliable energy resource

2 Give a reason why use of fossil fuels is being reduced.

3 Give an example of an unreliable energy resource.

Common misconception

Students may think that energy efficient appliances or systems are expensive. Emphasise that while some energy efficient appliances or systems may have a higher initial cost, due to advanced technology or special materials, they can save significant amounts of money in the long-term, so that the overall cost is lower for the owner.

Answers: 1 Any fossil fuel, biofuel, hydroelectricity, nuclear, geothermal *2* Any one from; non-renewable, running out, increase greenhouse gases, causing climate change *3* Any one from; wind, solar, wave

Trigger interest

Energy recap (10 mins)

In preparation for this debate, use the interactive to complete a series of paragraphs to summarise the basic principles of each energy resource studied in this chapter. Support students and develop reading and literacy by asking them to read sentences out loud, reinforcing their use of correct terminology.

Energy resources

Extend students by using the calculation sheet to interpret data about different energy sources, evaluate the data to make conclusions, and calculate the cost of energy.

Exposition

We need electricity (10 mins)

Set the scene by framing the discussion in terms of the energy needs of a modern society – specifically, in terms of electricity use. Discuss the implications of varying supply and demand, and the advantages of various resources in terms of reliability and start-up times. Students should be clear on the concept of a base-load and must be able to make comparisons of not only running costs but also initial (start-up) and final (decommissioning) costs to judge the overall cost effectiveness of the different resources.

Support students here by reminding them of the base-load graph and the power stations that are in constant operation to provide this power.

Main activity

The big energy debate (30 mins)

Contextualise this activity by referring to modern technologies that require large amounts of energy, e.g., AI, cryptocurrency, cloud gaming, and data centres. Students should appreciate the relevance of these examples and the criticisms they face because many of the companies that profit from them are often challenged because of adverse impacts on the environment.

Introduce a contentious statement or proposition to set up a debate, e.g., "The UK government should outlaw the use of AI, because it requires too much energy to run."

Remind students that the purpose of the debate is to reach a consensus (i.e., form conclusions) as a class, so they may create a set of proposals as required by the learning outcomes of the lesson.

Give timings and establish roles within the groups, if they are needed. To avoid a chaotic argument, discuss the ground rules for a successful discussion. Praise them for developing skills around research and fact-finding in the previous lesson, and encourage them to develop skills of waiting to speak, listening and persuading now.

Rules should include who can speak and when they can speak (a simple prop can help here, e.g., a small microphone or ball), who is recording discussions or decisions, and whether roles will be assigned or if students will be free to take on any position they want. Be mindful of the dynamics and the risk of gendered stereotyping – as far as possible, ensure all students are comfortable in the roles they assume, and that individuals or groups do not dominate inappropriately.

Review and practice

Democracy in action (5 mins)

Following the debate, provide students with a range of proposals for energy production that they must vote on. Make this an anonymous vote with ballot papers giving various options. The results can be declared immediately or during the next lesson.

As an alternative activity, students can write a letter to their MP outlining the decision they have reached in the big debate. They should explain what they have decided and ask their MP to act on the decision. This should be a formal letter, if possible.

P3 Energy resources

Introduction to checkpoint intervention

The checkpoint intervention provides suggestions for a lesson to follow up the *P3 Energy resources* checkpoint assessment. Depending on the outcomes of the assessment, these suggestions can be used to move a class towards secure, to target specific learning outcomes, or to consolidate knowledge for students achieving secure outcomes. Use students' outcomes from the checkpoint assessment to address general misconceptions from the content covered. Be prepared to re-cover content as required; three key concepts from the chapter are covered in more detail below.

Checkpoint secure learning outcomes

- Outline the operation of a nuclear power station.
- Describe the operation of a hydroelectric system.
- Explain why biofuels are considered carbon neutral.
- Suggest the most appropriate energy resource to use in a range of scenarios.
- Compare a wide range of energy resources in terms of advantages and disadvantages.
- Describe techniques to reduce the harmful products of burning fossil fuels.

Student reflection

Encourage students to reflect on whether there were any checkpoint questions they found difficult or straightforward as well as their preparation for the checkpoint assessment, including how they revised, what they revised, the time spent on revision, and what areas they could improve upon. For more strategies, see the Metacognition spread.

Intervention activity: Describe the operation of a hydroelectric system.

Ordering processes for hydroelectric systems

Ordering and describing of each process underpinning the production of electricity is challenging for students. This activity provides a framework so students can focus on ordering each stage of the process, followed by memory recall of each stage's descriptor.

Step-by-step

Create a simple card sort activity, ordered by number (on the back of the cards themselves, or hidden elsewhere), that describes how a hydroelectric system operates. There are five stages: collection of rainwater, storage in a reservoir, flowing downhill, driving turbines, generating electricity.

This activity works particularly well if students can match stages to a visual, for example, sticking cards to a diagram at each appropriate position of the flowing water's journey.

Intervention activity: Explain why biofuels are considered carbon neutral.

Biofuels are carbon neutral

Students are typically unaware of what carbon neutral actually means. They may be familiar with the idea of a carbon footprint, but context is key – the implication of increased CO_2 in the atmosphere (as a greenhouse gas) increasing global warming, and why we should all be aware of our carbon footprint and, if it is high, reduce it.

What's taken in is given back out

Provide a range of scenarios that demonstrate the idea of carbon offsetting, based on the idea of extracting and inserting CO_2 from and into the atmosphere, for example, airline campaigns about carbon-offsetting flights. Students can categorise which activity extracts CO_2, and which activity inserts CO_2 into the atmosphere. Explain that a carbon neutral process is one for which the amount in each case exactly balances – this is the key concept to link to the use of biofuels. The carbon that the living organism extracts from the atmosphere is, in theory, exactly balanced by the carbon reinserted once that organism is burned as a biofuel.

Intervention activity: Compare a wide range of energy sources in terms of advantages and disadvantages.

Energy production 'Snap'

Students are expected to learn a relatively large amount of information in order to compare a wide variety of energy resources. There are three learning stages here – identifying the energy resource (e.g., fossil fuels, nuclear power, and renewable sources), knowing their pros and cons, and then comparing them/weighing them up to arrive at a conclusion.

Step-by-step

Create a card game of 'snap' to help students to independently distinguish between the different energy resources, followed by an analysis of their pros and cons. You could, for instance, colour code a set of cards between fossil fuels (red), nuclear power (green) and renewable (light green). Once these cards are categorised, a simple discussion about pros and cons could follow – alternatively, have these listed as short bullet points on the back of each card.

2 Particles at work

Introduction to unit

Students will have already encountered particles, in simple terms, which will have informed them about density, diffusion, Brownian motion, and changes to internal energy when heated. Basic electrical concepts and circuits will also have been covered and these will be developed further in terms of current and static electricity. Finally, this unit covers radioactivity and the particles involved in nuclear fission, fusion, and radioactive decay.

Prerequisite knowledge from KS3 (chapter references from *Oxford Smart* Activate)

P2.1.2 Circuits and current

- Electric current, measured in amperes, series and parallel circuits.

P3.4.4 Apply: Current, p.d., and resistance

- Potential difference, measured in volts, battery and bulb ratings; resistance, measured in ohms, as the ratio of potential difference (p.d.) to current.
- Differences in resistance between conducting and insulating components (quantitative).

P2.3.3 Pressure in gases, P2.3.4 Pressure in liquids, P2.3.5 Pressure on solids

- Diffusion in liquids and gases driven by differences in concentration.
- The difference between chemical and physical changes.
- The differences in arrangements, in motion and in closeness of particles explain changes of state, shape and density, the anomaly of ice-water transition.

Assessment

Suggested minimum guided learning hours: 27 hours

Exam: Paper 1

AQA GCSE Physics specification links

4.2 Electricity

- 4.2.1 Current, potential difference, and resistance
- 4.2.2 Series and parallel circuits
- 4.2.3 Domestic uses and safety
- 4.2.4 Energy transfers

4.3 Particle model of matter

- 4.3.1 Changes of state and the particle model
- 4.3.2 Internal energy and energy transfers
- 4.3.3 Particle model and pressure

4.4 Atomic structure

- 4.4.1 Atoms and isotopes
- 4.4.2 Atoms and nuclear radiation
- 4.4.3 Hazards and uses of radioactive emissions and background radiation (physics only)
- 4.4.4 Nuclear fission and fusion (physics only)

Identity, awe, and wonder

Students will be aware that physics deals with the behaviour of nature at a very large scale (e.g., the Universe), which we call the macroscopic level and at the very small scale (e.g., the atom), which we call the microscopic level. They are likely to be fascinated by the fact that incredibly tiny particles, such as electrons, enable energy to be transferred electrically to power whole cities. Similarly, the tiny neutrons that are absorbed by uranium nuclei also provide enormous amounts of energy to enable electricity to be generated in nuclear power stations. Nuclear radiation is often seen as very dangerous, but the process of nuclear fission provides enormous energy per kilogram of fuel and the radiation from radioactive sources can be very beneficial to humans. Students could investigate the advantages and disadvantages of nuclear physics and evaluate whether it is a good thing or a bad thing.

Required practicals

AQA required practicals	Practical skills	Topic
Required practical 3: Resistance Use circuit diagrams to set up and check appropriate circuits to investigate the factors affecting the resistance of electrical circuits. This should include: • the length of a wire at constant temperature • combinations of resistors in series and parallel.	AT 1 – use appropriate apparatus to measure and record length accurately. AT 6 – use appropriate apparatus to measure current, potential difference, and resistance. AT 7 – use circuit diagrams to construct and check series and parallel circuits.	P4.3; P4.6
Required practical 4: I-V characteristics Use circuit diagrams to construct appropriate circuits to investigate the I–V characteristics of a variety of circuit elements including a filament lamp, a diode, and a resistor at constant temperature.	AT 6 – use appropriate apparatus to measure current and potential difference and to explore the characteristics of a variety of circuit elements. AT 7 – use circuit diagrams to construct and check series and parallel circuits including a variety of common circuit elements.	P4.4
Required practical 5: Density Use appropriate apparatus to make and record the measurements needed to determine the densities of regular and irregular solid objects and liquids. Volume should be determined from the dimensions of a regularly shaped object and by a displacement technique for irregularly shaped objects. Dimensions to be measured using appropriate apparatus such as a ruler, micrometer or Vernier callipers.	AT 1 – use appropriate apparatus to make and record measurements of length, area, mass and volume accurately. Use such measurements to determine the density of solid objects and liquids.	P6.1

Working scientifically

Students will understand how scientific methods and theories have developed over time, giving us a better understanding of the physical world based on the most up to date and reliable data, collected from experiments, taking into consideration the importance of peer review. They will also use a variety of models to make predictions and draw conclusions from their results and observations.

Students will consider the ethical aspects of science and use data to evaluate the problems caused by human impacts on the environment. Students will be expected to use correct scientific vocabulary, quantities, and units when communicating their knowledge.

Maths skills

Students will be expected to recognise and use fractions, decimals, and percentages in calculations, and represent very large and small numbers in standard form. They will rearrange formulae and substitute values in order to calculate the size of physical quantities, using appropriate units.

Students will also plot graphs and interpret data from these graphs in numerical form.

Literacy links

The practical investigations that students will encounter in this unit will serve as an opportunity for writing skills, such as the writing of a step-by-step method. Other literacy-based activities in this unit can be found in P8.1, P8.5, P9.4, and P10.4.

Physics only

P4 Electric circuits

P4.1 Electrical charges and fields

AQA GCSE Physics specification points

4.2.5.1 Static charge

4.2.5.2 Electric fields

Smart Curriculum Narrative

Electricity and magnetism

Theme/subtheme

Electricity and magnetism/Charge

Disciplinary knowledge

- WS 1.2, 1.5

Practical

Charging by friction

	Learning outcomes		
Learning objectives	**Developing Aiming for 3**	**Secure Aiming for 5**	**Extending Aiming for 8**
Use the concept of electric fields to explain why charged objects interact	Draw the electric field pattern around a point charge or charged sphere	**Use the concept of electric fields to explain why charged objects interact**	Apply the concept of electric fields to explain in detail why the force between charged objects decreases with increasing distance
Describe evidence that forces between charged objects act a distance	Describe the interactions between positively and negatively charged objects	**Describe evidence that forces between charged objects act a distance**	Explain how charged objects can attract neutral objects
Describe how objects become negatively or positively charged in terms of electron transfer	State that objects can become electrically charged by the action of frictional forces	**Describe how objects become negatively or positively charged in terms of electron transfer**	Explain why sparks can be produced by charged materials in terms of charge build-up

Tier 2 vocabulary	Tier 3 vocabulary
atom, *current*, electricity, electron, force, friction, ion, negative, positive	attraction, *charge*, discharge, electrostatic, field lines, force field, repulsion, *static*

Lesson plan

Reactivate knowledge

1 Define a non-contact force.

2 What happens when negatively charged objects come close to each other?

3 What is an electric field?

Common misconception

Students may think that static electricity is the spark observed during an electric shock. Explain that it is simply an imbalance of charge within or on the surface of or between materials. These charges are not 'moving'; motion of charged particles only occurs during sparking, which is a discharge of electric current.

Answers: 1 A force that acts when the objects are at a distance *2* Repel *3* An area around a charged object

Trigger interest

Flying saucers (5 mins)

Place four small metal pie tins on top of a Van de Graaff generator (VDG). Ask students to predict what will happen once the VDG is switched on, then demonstrate the tins flying off as the VDG charges.

Encourage students to verbalise what they observe; provide a list of keywords (e.g., electrons, repel, negative charge, friction) for additional support. Alternative demonstrations are provided on the practical sheet.

Exposition

The atom (20 mins)

Recap the structure of an atom, using a visual prompt, ensuring that students understand how electrons can be removed from atoms to form ions. Video, or a suitable animation, works well here – transferral of electrons between materials is the first key concept.

Invisible force fields

The second key concept of this lesson relates to force fields, which exist around charged objects (hence, this idea should follow learning about how objects become charged by transfer of electrons).

Demonstrate the idea of a field using simple bar magnets, leading students to conclude that non-contact forces are exerted because of an invisible magnetic field around them. Then, by referring back to the VDG, discuss the presence of an invisible electric field around the dome: electrically charged objects brought near to it would experience a force. Draw field lines around a simple diagram of the charged sphere, showing the direction that a positively charged object will be pushed.

Main activity

Charging by friction (20 mins)

Students should use the practical to explore motion of electrons between materials – the mechanism for this movement being frictional forces.

As an extension exercise, students could explore a series of other simple electrostatics experiments in a circus activity, for example, observing how charged balloons 'stick' to a cupboard door or how a trickle of water from a tap bends in response to a charged perspex rod brought close to it. Encourage students to explain these effects in terms of the movement of electrons, attraction/repulsion, and electric fields.

Circulate, reminding students that charged objects have an electric field around them which exerts a non-contact force. Expect that this concept will not be as intuitively grasped as charging by friction.

Review and practice

Static force (10 mins)

Use the interactive to summarise the key ideas about charged objects, transferral of electrons, and electric fields. Show students a set of diagrams of charged objects with force arrows (of attraction and/or repulsion) drawn. Ask students to assess which are correct and which are incorrect.

P4.2 Current and charge

AQA GCSE Physics specification points

4.2.1.1 Standard circuit diagram symbols

4.2.1.2 Electrical charge and current

Smart Curriculum Narrative

Electricity and magnetism

Theme/subtheme

Electricity and magnetism/Circuits

Disciplinary knowledge

- WS 1.2
- MS 3a, 3b, 3c

Activity

Circuit tests

	Learning outcomes		
Learning objectives	**Developing** **Aiming for 3**	**Secure** **Aiming for 5**	**Extending** **Aiming for 8**
Draw and interpret simple circuit diagrams	Identify circuit components from their symbols	**Draw and interpret simple circuit diagrams**	Design a circuit to meet a requirement
Describe what current is and how it is measured	State the definition for current and its unit	**Describe what current is and how it is measured**	Explain why a bulb lights immediately a switch is pressed
Calculate the charge flow by a steady current in a given time	State that the charge flow through a component depends on the current and the time	**Calculate the charge flow by a steady current in a given time**	Perform a range of calculations, including rearrangement of the equation $Q = It$
Construct an electrical circuit and accurately measure the current	Construct a simple electrical circuit	**Construct an electrical circuit and accurately measure the current**	Measure the current in a circuit accurately and use it to calculate the rate of flow of electrons

Tier 2 vocabulary	Tier 3 vocabulary
charge, *current*, electric, flow, variation, wire	ammeter, component, potential difference, resistor

Lesson plan

Reactivate knowledge

1 What electrical components make up a torch?

2 What instrument measures current?

3 What is an electric current?

Common misconception

Students may often think that circuit symbols are meaningless. Respond that each circuit symbol has historical background and rationale which can be used as an extension activity or a discussion. For example, the arrow used for a variable resistor alludes to the dial of an analogue meter, which were once commonly used to measure potential difference and current variations in labs.

Answers: 1 Battery/cell, lamp, switch, leads *2* Ammeter *3* The amount of charge flowing per second

Trigger interest

It's symbolic (10 mins)

Show a set of slides/diagrams to students containing a mix of common symbols (which they interpret) or words (which they draw the symbol for) on their mini-whiteboard. Use common road signs, hazard symbols, brands, common music, and computer symbols (e.g., play, pause, save, search).

Link this activity to circuit symbols that are used to represent components in electric circuits. Continue the activity with circuit symbols themselves to test prior knowledge. Support students by drawing and building a simple circuit in tandem to emphasise the use of straight lines as connecting wires and symbols as circuit components.

Exposition

Driving current (15 mins)

Introduce the idea of a flow of current using a suitable video/animation. PhET demos are useful, especially if virtually building the circuits in tandem with the real wires and components. Expect students to grasp that current is represented by flowing electric charge (which can be seen explicitly using the animated demo), and that a quicker flow (rate of charge flow) represents a larger current.

Demonstrate this using the virtual ammeter and a real one. Draw attention to the fact that once a circuit is complete and connected, all free charges around the circuit start to move at the same time (a misconception is that charges only start moving at the power source). From this, lead students to the defining equation:

current = charge flowing per unit of time, or $I = \frac{Q}{t}$, which rearranged is $Q = It$

A few simple calculations are useful to embed the meaning of I and Q, and their units.

Ask students to identify which component is responsible for pushing current around the circuit. Refer to this as a source of potential difference, which will be covered conceptually in the next lesson.

Main activity

Circuit tests (25 mins)

Allow students to begin independently but offer support in a different area in the classroom, if available, if students need additional help setting up their circuits.

Demonstrate how a circuit should be constructed methodically, for example, which components can remain and which can be swapped in and out as desired. Additional support can be provided by providing students with a photograph of the equipment set up, alongside circuit diagrams.

Review and practice

Current calculations (5 mins)

Using mini-whiteboard, give students some basic calculations to solve using the equation.

Extend students' understanding by using unit prefixes like mA or units of time in minutes and hours.

P4.3 Potential difference and resistance

AQA GCSE Physics specification points

4.2.1.3 Current, resistance, and potential difference

Smart Curriculum Narrative

Electricity and magnetism

Theme/subtheme

Electricity and magnetism/Currents, resistance, and potential difference

Disciplinary knowledge

- WS 2.3, 2.4, 2.5, 2.6, 2.7, 3.5, 3.6, 3.7, 3.8, 4.2, 4.3, 4.6
- MS 3b, 3c

Required Practical 3

Resistance

	Learning outcomes		
Learning objectives	**Developing Aiming for 3**	**Secure Aiming for 5**	**Extending Aiming for 8**
Use the equation that links potential difference, energy transferred and charge to calculate potential difference	State the definition of potential difference and its unit	**Use the equation that links potential difference, energy transferred and charge to calculate potential difference**	Rearrange equation for potential difference
Calculate the resistance of a component	State that resistance restricts the size of a current in a circuit	**Calculate the resistance of a component**	Rearrange equation for resistance
Describe how the resistance of thermistors and LDRs changes	State that the resistance of thermistors and LDRs is not constant	**Describe how the resistance of thermistors and LDRs changes**	Describe how LDRs and thermistors can be used in sensing circuits
Measure the effect of changing the length of a wire on its resistance in a controlled experiment	Measure the current and potential difference in a circuit to determine the resistance	**Measure the effect of changing the length of a wire on its resistance in a controlled experiment**	Devise an investigation to measure the effects of length of wire on resistance focussing on quality of data

Tier 2 vocabulary	Tier 3 vocabulary
charge, component, coulomb, electron, flow, force, joule, transfer, voltage	circuit, conductor, current, potential difference (p.d.), resistance, voltmeter

Lesson plan

Reactivate knowledge

1 What is potential difference?
2 How do you measure the potential difference?
3 What determines the current in a circuit?

Common misconception

Students may think voltage/p.d. is a force. Respond that p.d. is the amount of energy transferred per unit of charge as a current flows across a component. Emphasise energy transfer; a volt is equivalent to one joule per coulomb so, a 6V battery transfers 6 joules per coulomb of charge.

Answers: 1 Energy transferred per unit charge *2* Voltmeter *3* The p.d. of the supply and the resistance of the components

Trigger interest

Resisting motion (5 mins)

Frame this lesson as a story about the journey of electrons through a component. This journey is fraught with risk of energy transfer: firstly, due to crossing a p.d. and secondly due to collisions. These two ideas, and their associated equations, need to be carefully distinguished.

Use volunteers to illustrate the effect of resistance; this can be done using desks to block the paths of the students as they walk swiftly through, or by using students themselves to obstruct the path as 'vibrating ions'. Illustrate the point more accurately by showing students an animation or video of electrons drifting through a metal lattice, drawing out ideas about energy transfer and collisions.

Exposition

Resistance depends on length (15 mins)

Discuss the nature of p.d., using analogies as appropriate, to emphasise it as an energy transfer. Pens rolling off a sloping mini-whiteboard works nicely as a simple practical analogy to show that a larger p.d. (i.e., larger slope) drives more current (rate of charge flow).

Show how a p.d. is measured in a circuit, using a voltmeter, again emphasising the interplay between volts and joules per coulomb so students link this measurement to that of an energy transfer. Introduce the equation in words, then use the Maths skills interactive to give students some practice applying it to simple examples.

Gently introduce the idea that resistance is a ratio of p.d. to current, by revisiting the narrative of the lesson of the electron journey through a component.

Given that they have to make their way through an array of 'large' vibrating metal ions, the amount of resistance to their motion is expressed as:

$$R = \frac{V}{I} \quad \text{and so} \quad V = IR$$

Students should be able to recognise and apply this equation in both forms. Reinforce the important role of vibrating metal ions, linking this to the temperature of the metal wire and the fact that Ohm's law requires this to be constant.

Main activity

Resistance (25 mins)

In this required practical, different groups of students can be given wires of different diameters or materials to show that the pattern is the same (resistance is proportional to length) and identify some of the additional factors that might affect resistance (e.g., cross-sectional area).

Support students by providing blank results tables. As you circulate, extend students' conceptual understanding of resistance by asking them to think about the journey of electrons through their wires and the interactions between these and the positively charged metal ions.

Review and practice

A proportionate relationship (10 mins)

Scaffold questions about potential difference and resistance in a mini-whiteboard quiz. Lead students to appreciate the mathematical relationship between length and resistance, that is, that resistance is proportional to length. Additional calculations, using the equation, should be used to reinforce learning and emphasise units, differentiated by student ability as appropriate.

If there is time, extend students by drawing out ideas about which of the particles in a metal wire are in fixed positions, and which are free to move, their relative sizes, and the exchange of energy between kinetic and thermal stores during collisions. Hence, electrical resistance leads an increasing temperature!

P4.4 Component characteristics

AQA GCSE Physics specification points

4.2.1.4 Resistors

Smart Curriculum Narrative

Electricity and magnetism

Theme/subtheme

Electricity and magnetism/Currents, resistance, and potential difference

Disciplinary knowledge

- WS 1.2, 1.4
- MS 4c, 4d, 4e

Required Practical 4

I–V characteristics

Learning objectives	Learning outcomes: Developing Aiming for 3	Learning outcomes: Secure Aiming for 5	Learning outcomes: Extending Aiming for 8
Using positive and negative axes and choosing suitable scales plot *I–V* graphs for components	Measure the current and potential difference across a lamp and plot *I–V* graphs	**Using positive and negative axes and choosing suitable scales plot *I–V* graphs for components**	Determine the resistance of a component based on information extracted from a *I–V* graph
Describe how the resistance of resistors, diodes and filament lamps changes	Identify components from simple *I–V* graphs	**Describe how the resistance of resistors, diodes, and filament lamps changes**	Explain the resistance characteristics of a filament lamp in terms of electrons and ion collisions
Explain how you can identify an ohmic conductor from *I–V* graph	State what is meant by an ohmic conductor	**Explain how you can identify an ohmic conductor from *I–V* graph**	Use the gradient of an *I–V* graph to compare the resistance of different components

Tier 2 vocabulary	Tier 3 vocabulary
component, dependent variable, independent variable, reading, temperature	ammeter, diode, light dependent resistor, parallel, resistance, series, thermistor, variable resistor, voltmeter

Lesson plan

Reactivate knowledge

1 How do you measure the resistance of a component?
2 How do you calculate the resistance of a component?
3 Name a component whose resistance varies?

Common misconception

Students may think that resistance depends on potential difference and current because of Ohm's law. Tell students resistance is only dependent on the length of conductor, its cross-sectional area, and the type of conducting material.

Answers: 1 Connecting an ammeter in series and a voltmeter in parallel *2* Divide the p.d. by the current *3* Any from: variable resistor/ LDR/thermistor/filament lamp/diode

Trigger interest

Characteristic fingerprints (5 mins)

Discuss the use of fingerprints to uniquely identify people, and contrast this with how physicists use current and potential difference to uniquely identify different circuit components. Remind students that the ratio of p.d. to current defines the resistance of a component, and this quantity behaves in different ways depending on the component itself. Introduce the three components that students will investigate for the lesson: a fixed resistor, a filament lamp, and a diode.

Exposition

Investigating thermistors and light dependent resistors (LDRs) (10 mins)

Introduce two 'suspects' under investigation and re-emphasise (as much as possible in this lesson!) that to investigate their characteristics, the current should be measured as the p.d. across them changes, alluding to the variables in the practical.

Prepare a circuit in advance with a thermistor attached by crocodile clips. The point here is not the setup of the circuit – rather, focus on the collection of a pair of values from the ammeter and voltmeter for both the thermistor, and then the substituted LDR. Firstly, lower the thermistor carefully into a beaker of hot water and ask two volunteers to observe what happens on the ammeter and voltmeter. Draw out two conclusions: (a) the current and p.d. is changing, so the resistance must be changing, and (b) the value of resistance at some instant. It will be useful to have the equation displayed, ready to substitute values. Demonstrate collection of pairs of values in a similar way for the LDR.

Main activity

I–V characteristics (30 mins)

Start by discussing the circuit setup for investigating wire characteristics from P4.3. Ensure students understand the placement of an ammeter in series and a voltmeter in parallel, along with the roles of a switch and variable resistor. Emphasise that these elements are standard in any circuit used for component investigations. Draw a large diagram on a white board, leaving spaces for swappable components (fixed resistor, filament lamp, diode). Differentiate circuit symbols with colour or sticky notes to highlight component interchangeability while maintaining the basic circuit structure.

Remind students of the appropriate supply p.d. to use. Ensure they start their measurements when the variable resistor is at its lowest resistance (so current is at its highest, but below 1.0 A). As students begin taking readings, circulate and provide support where necessary, for example, with a template results table. For some groups, it may be appropriate to only investigate one component, and share their results with others.

Display three I–V axes on a whiteboard for students finishing first to sketch the shape of their line, providing the basis for class-wide progress assessment and discussion.

Review and practice

Thermistors and LDRs (10 mins)

Revisit the circuit from the start of the lesson, with the same volunteers. Ask them to take a pair of readings and calculate resistance in the same way. Expect all students to immediately conclude that the resistance has increased (not just changed).

Scaffold questions to get the conclusion that temperature affects the resistance of a thermistor, and that the resistance increases as the temperature decreases. Extend students' understanding by asking them to sketch a graph of resistance against temperature.

Finally, ask them to predict what might happen to the resistance of an LDR if it was placed close to a bright light. Set up the circuit to demonstrate.

P4.5 Series circuits

AQA GCSE Physics specification points

4.2.2 Series and parallel circuits

Smart Curriculum Narrative

Electricity and magnetism

Theme/subtheme

Electricity and magnetism/Circuits

Disciplinary knowledge

- WS 1.4
- MS 1c, 3b, 3c, 3d

Practical

Investigating potential differences in a series circuit

	Learning outcomes		
Learning objectives	**Developing Aiming for 3**	**Secure Aiming for 5**	**Extending Aiming for 8**
State that the current in any part of a series circuit is the same	State that there is only one loop in a series circuit	**State that the current in any part of a series circuit is the same**	Explain in detail why the current in a series circuit is the same at all points by using the concept of conservation of charge (electrons)
Investigate and describe how potential difference across components in series relates to potential difference of the power supply	Calculate the potential difference provided by cell combinations	**Investigate and describe how potential difference across components in series relates to potential difference of the power supply**	Evaluate in detail the investigation of series circuits and explain discrepancies
Calculate the total resistances of two resistors in series and explain why total resistance increases	State that the total resistance increases when you add more components in series	**Calculate the total resistances of two resistors in series and explain why total resistance increases**	Analyse a variety of series circuits to determine the current through, p.d. across, and resistance of combinations of components

Tier 2 vocabulary	Tier 3 vocabulary
component, conservation, *charge*, dissipation, electric, energy, flow, thermal	ammeter, current, internal resistance, loop, ohmmeter, parallel, p.d., series, voltmeter

Lesson plan

Reactivate knowledge

1 What is a series circuit?

2 Does the current increase/decrease/stay the same in all parts of a series circuit?

3 What happens to the current in a series circuit if you add more components?

Answers: 1 One where there is only one loop *2* Stay the same *3* Current decreases

Common misconception

Students may think electrons and/or current is used up in a component. Respond that current is the flow of electric charge through a circuit which cannot be used up in a component, only transferred. The amount of energy transfer (per unit of charge) is called the p.d. across the component.

Trigger interest

Simulating current in series (10 mins)

Use simulations (PhET circuit builders are great) as much as possible in this lesson to enable students to visualise the flowing charge around a series circuit. Direct observation should hint that the rate of flow is the same at each point in the loop.

Build on this observation using multiple virtual ammeters at three different points to illustrate this idea quantitatively. Students should immediately conclude that current at all points in a series 'loop' is equal.

Extend students by exploring the common misconception that current gets used up in a component: explore ideas for what is really used up, or more accurately, transferred as current flows through a component. Allude to the idea of conservation – students should be able to see that the amount of charge going into a component is equal to the amount leaving it.

Exposition

One-way systems (10 mins)

Set up the hinge question, to be revisited at the end of the lesson: In what situation are objects only allowed one-way through something? Encourage students to provide some suggestions (e.g., one-way traffic on roads, people through a ride at an amusement park, liquid going down a sink), and to keep this idea in mind as they explore the properties of series circuits in the practical activity.

Main activity

Investigating potential differences in a series circuit (30 mins)

Demonstrate the conservation of current using a real circuit, ideally an identical circuit to that used in the virtual demo. Next, demonstrate how to explore p.d.'s using a series of voltmeters: a student can do this in tandem using the virtual simulation.

Pairs of students use the practical to investigate p.d. of components in a loop, concluding that these sum to the p.d. provided by the power supply. There will be some variation in measurements – the p.d.'s across each component should sum to slightly less than the p.d. provided by the power supply. A discussion of meter precision and random error should take place.

Resistors in series (demonstration)

Call small groups up to a station in the classroom, away from their practical investigation. Using a simple circuit, set up using PhET and an ohmmeter to see values of resistance directly, students should test the given result that the total resistance is the sum of each individual resistance.

Review and practice

One-way systems (5 mins)

Develop analogies for current flowing in a series circuit, based on student ideas about one-way systems. It is important to distinguish each of the three ideas covered in this lesson: current is the same at each point in a series loop, p.d.'s are shared, and resistances add to give a total resistance.

Extend student understanding by using the Interactive: *Controlling current* to explore combinations of cells and resistors to give certain values of current.

P4.6 Parallel circuits

AQA GCSE Physics specification points

4.2.2 Series and parallel circuits

4.2.1.3 Current, resistance, and potential difference

Smart Curriculum Narrative

Electricity and magnetism

Theme/subtheme

Electricity and magnetism/Circuits

Disciplinary knowledge

- WS 1.4
- MS 1c, 3b, 3c, 3d

Required Practical 3

Resistance

	Learning outcomes		
Learning objectives	**Developing Aiming for 3**	**Secure Aiming for 5**	**Extending Aiming for 8**
State that the total current in a parallel circuit is the sum of the current through each loop of the circuit	Describe a parallel circuit as one with more than one loop	**State that the total current in a parallel circuit is the sum of the current through each loop of the circuit**	Determine the current at any point in a parallel circuit
Investigate the effect of adding resistors in parallel on the size of the current in a circuit and explain why total resistance is less	State the effect of adding resistors in parallel on the size of the total resistance of the circuit	**Investigate the effect of adding resistors in parallel on the size of the current in a circuit and explain why total resistance is less**	Evaluate in detail an investigation into the effect of adding resistors in parallel on a circuit
Measure the p.d. across parallel circuits and explain any discrepancies	State that the p.d. across parallel sections of a circuit is the same	**Measure the p.d. across parallel circuits and explain any discrepancies**	Analyse a variety of parallel circuits to determine missing currents, potential differences, and calculating resistance of components

Tier 2 vocabulary	Tier 3 vocabulary
battery, flow, kilogram, mass, second	ammeter, circuit, *current*, parallel, potential difference, resistance, resistor, series, voltmeter

Lesson plan

Reactivate knowledge

1 What is a parallel circuit?

2 What is potential difference?

3 Compare the potential difference in bulbs.

Common misconception

Students may think that adding resistors to a circuit always increases the total resistance. Tell students that resistors added in a series circuit does increase the total resistance in the loop. Adding a resistor in parallel decreases the total resistance across that branch.

Answers: 1 Has more than one loop *2* Energy transferred per unit charge *3* They have the same p.d.

Trigger interest

The river (5 mins)

Show students an image of a river branching and rejoining. Link this image to analogies discussed in the previous lesson about current flow. Annotate points on the image before, during, and after the split and ask students to use numbers to illustrate what happens to the river current (mass of water passing each point per second) at each point, in units of kilograms per second.

Support students by providing some indicative values; they could use mini-whiteboards to intuitively deduce missing current values along each branch or vote which set of values show the correct combination of currents.

Exposition

Simulating current in parallel (10 mins)

As in the previous lesson, use a simulation in tandem with a real circuit to investigate parallel currents and potential differences. Students may find parallel circuits trickier to set up, therefore have these prepared in advance so that they can focus on the current and p.d. rules, at this stage of the lesson. Get volunteers to place ammeters and voltmeters at appropriate positions to spot patterns.

Measure the current in two branches and allude to what the total current must be (allowing students to confirm their intuition in the practical activity). Then, show explicitly that potential difference across parallel branches is the same. Students may need additional scaffolding identifying a 'branch', and then a 'parallel branch', so make sure this is explicitly shown.

Main activity

Resistance (30 mins)

The first half of the required practical is covered in P4.3.

Careful supervision may be required while students build parallel circuits. A recommended approach is to build them loop-by-loop, such that the first complete loop is the one containing the power supply. Once students can identify and build this loop, they should prepare separate branches of the circuit, connecting these branching loops wherever required.

Direct students to confirm their intuition about current in separate branches using an ammeter to measure the current in the two branches as well as the total current. As you circulate, emphasise measurement of the total current near the terminals of the power supply, and check that students can see this is the sum of current in each branch. Remind them this will not be exact due to meter errors.

Once students feel confident with building and analysing their parallel circuit, they should analyse the circuit using the practical resource to consolidate the current and p.d. rules.

Review and practice

Another circuit (10 mins)

Use the interactive where students analyse a parallel circuit to determine the current from a battery.

P4 Electric circuits

Introduction to checkpoint intervention

The checkpoint intervention provides suggestions for a lesson to follow up the *P4 Electric circuits* checkpoint assessment. Depending on the outcomes of the assessment, these suggestions can be used to move a class towards secure, to target specific learning outcomes, or to consolidate knowledge for students achieving secure outcomes. Use students' outcomes from the checkpoint assessment to address general misconceptions from the content covered. Be prepared to re-cover content as required; three key concepts from the chapter are covered in more detail below.

Checkpoint secure learning outcomes

- Describe how objects become negatively or positively charged in terms of electron transfer.
- Use the equation that links potential difference, energy transferred, and charge to calculate potential difference.
- Calculate the resistance of a component.
- Describe how the resistance of resistors, diodes, and filament lamps changes.
- Measure the effect of changing the length of a wire on its resistance in a controlled experiment.
- State that the current in any part of a series circuit is the same.

Student reflection

Encourage students to reflect on whether there were any checkpoint questions they found difficult or straightforward as well as their preparation for the checkpoint assessment, including how they revised, what they revised, the time spent on revision, and what areas they could improve upon. For more strategies, see the Metacognition spread.

Intervention activity: Use the equation that links potential difference, energy transferred, and charge to calculate potential difference.

Understanding potential difference

This is a very tricky concept to fully understand at GCSE, especially considering the multiple terms for the same concept (voltage, p.d., potential difference). Students are yet to understand the meaning of a potential, or a full idea of electric fields, so focus on using simple analogies that emphasise an energy transfer, and direct applications of the equation.

Using the equation

Collect a sample of cells and power sources and ask students to translate the given potential differences in terms of joules per coulomb. Draw diagrams of whole coulombs of charge – which contains many individual electrons! – to emphasise the quantity of energy transferred by each unit coulomb. Scaffold these questions into simple calculations that use the equation more directly, emphasising units along the way. An equation triangle is useful support for students, where needed.

Intervention activity: Describe how the resistance of resistors, diodes, and filament lamps changes.

Using current to interpret changes in resistance

Encourage students to focus on describing how the current changes through each of these components as the p.d. increases. This can be done experimentally (e.g., by focussing on an ammeter reading) or graphically. Students should apply the general principle that large resistance impedes current flow, so any indication that current is small (or being impeded) will indicate that resistance is large (or increasing).

Three students, three mystery components

Create a short story about three students who each record two pairs of readings from a voltmeter and an ammeter, dividing the pairs to calculate resistance. Describe how the calculated value changes for each student; this information can be used to determine which component is being investigated. This can be scaffolded as appropriate – with small groups to make the investigative activity more competitive. Provide 'clues' where needed for additional support, for example, a mini-whiteboard with a sketch of each *I-V* graph.

Intervention activity: Measure the effect of changing the length of wire on its resistance in a controlled experiment.

Understanding potential difference

As an intervention activity, ensure wherever possible that the circuit for this experiment is prepared in advance to enable students to focus on measuring the desired effect. This is that resistance increases with length.

Using an ohmmeter

A multimeter, used as an ohmmeter, can be used to directly observe that the resistance increases with increasing length of wire in the circuit. Alternatively, a simulation will suffice; spend time ensuring students understand this intuitively (flowing charge encounters more collisions, and hence more resistance, for a greater length of wire). Extend this experimental activity by using an ammeter and a voltmeter, emphasising that pairs of readings are required to determine resistance, which is expected to increase with length.

P5 Electricity in the home

P5.1 Alternating current

AQA GCSE Physics specification points

4.2.3.1 Direct and alternating potential difference

Smart Curriculum Narrative

Electricity and magnetism

Theme/subtheme

Electricity and magnetism/Domestic electricity

Disciplinary knowledge

- MS 4a

Practical

Investigating an alternating potential difference

	Learning outcomes		
Learning objectives	**Developing Aiming for 3**	**Secure Aiming for 5**	**Extending Aiming for 8**
Describe the characteristics of the UK mains supply	State that the frequency and potential difference of the UK mains supply	**Describe the characteristics of the UK mains supply**	Suggest why travel adaptors are needed when visiting other countries
Describe the difference between direct and alternating voltage	State examples of sources of alternating and direct voltage supplies	**Describe the difference between direct and alternating voltage**	Analyse a.c. traces from an oscilloscope to determine the voltage and frequency
Explain why the National Grid is an efficient system to transfer energy	Describe the National Grid	**Explain why the National Grid is an efficient system to transfer energy**	Explain why the National Grid is needed

Tier 2 vocabulary	Tier 3 vocabulary
dissipate, electron, energy, frequency, powers, resistance, symbol, transfer, voltage, wave, wire	alternating current, circuit, direct current, efficiency, mains supply, National Grid, oscillate, oscilloscope, peak, *trace*, transform

Lesson plan

Reactivate knowledge

1 What is current?

2 What is potential difference?

3 How are current and potential difference related?

Common misconception

Students may think that the potential difference of d.c. supply is always constant. Respond that d.c. can vary. Sometimes, it's confused with a.c, which also varies; but the key difference is that the polarity of d.c. does not change (i.e., the positive and negative terminals do not swap around). In other words, the potential difference of d.c. might vary positively or negatively. However, the potential difference of a.c. supplies vary continuously from positive to negative.

Answers: 1 Rate of flow of charge *2* Energy transferred per unit charge *3* Current is directly proportional to the potential difference

Trigger interest

Comparing alternating current/direct current (5 mins)

Use the PhET circuit construction kit to create two simple direct current (d.c.) and alternating current (a.c.) circuits side by side (two screens may be required). Students should compare and contrast the circuits, noting similarities and differences, including the circuit symbols and motion of the electrons.

Support students by providing keywords/phrases to use in their observations, such as 'oscillate' and 'one-direction'.

Note that the PhET symbols for lamp and resistor is not used by AQA.

Exposition

Role playing electron movement (15 mins)

If space permits, use the whole group or small groups of students to roleplay the movement of electrons in a d.c. circuit and an a.c. circuit. Emphasise the idea that in both cases, electrons transfer energy to devices, and in particular that a.c. supplies, via mains circuits, power most homes in the UK.

Introduce the idea that, in an a.c. circuit, the frequency of oscillation is an important property, linking this into the frequency of mains supply.

Emphasise the higher, rapidly varying potential differences. The characteristics of mains (50 Hz and around 230 V) should also be covered, noting that peak potential difference is significantly higher (325 V).

Outline the basic features of the National Grid in terms of transformers and changes in voltages, using images where appropriate. Challenge students, where appropriate, to justify ideas about power efficiency at high potential differences, with reference to the equation for power, $P = VI$.

They should be able to recall, from prior learning, that large currents dissipate energy due to electrical resistance.

Main activity

Demonstration: Introducing oscilloscopes (5 mins)

Demonstrate the use of an oscilloscope to visualise oscillating signals, such as an a.c. supply.

Investigating an alternating potential difference (20 mins)

Wherever possible for this practical, encourage teaching assistants and/or technicians to support with setup in advance. The key learning is to ensure that the students can distinguish a.c. and d.c. signals and go further to interpret them in terms of peak potential difference and time period.

Review and practice

Mains facts (10 mins)

The relevance and prevalence of a.c. should be made clear to students. Students are likely to interact with mains powered (i.e., alternating current) every day; they should also appreciate that historically, the adoption of a.c. had a profound impact on global society, technology and the economy - in effect, because it enabled electricity to be efficiently distributed over vast distances, 'electrifying' large countries and industries. Students could imagine what everyday life would be like without some common household appliances, which has been made commonplace in the UK because of electricity distribution via a.c. by the National Grid. Use the interactive to quiz students about the key facts from the lesson.

P5.2 Cables and plugs

AQA GCSE Physics specification points

4.2.3.2 Mains electricity

Smart Curriculum Narrative

Electricity and magnetism

Theme/subtheme

Electricity and magnetism/Domestic electricity

Disciplinary knowledge

- WS 1.5

Activity

Plugs, cables, and short circuits

Learning objectives	Learning outcomes Developing Aiming for 3	Secure Aiming for 5	Extending Aiming for 8
Describe the functions of the three wires in a three core cable	Identify the live, neutral, and earth wires in a three-pin plug	**Describe the functions of the three wires in a three core cable**	Explain when there will be a current in the live, neutral, and earth wires of an appliance
Describe why a short circuit inside a device presents a hazard	Identify simple and obvious hazards in electrical wiring	**Describe why a short circuit inside a device presents a hazard**	Explain why you should never touch the live wire
Discuss the choices of materials used in cables and plugs in terms of their physical and electrical properties	Identify the key components of a typical three-pin plug and socket	**Discuss the choices of materials used in cables and plugs in terms of their physical and electrical properties**	Explain why the fuse is connected to the live wires

Tier 2 vocabulary	Tier 3 vocabulary
battery, bulb, collision, crocodile clip, electric, flammable, flow, heating, parallel, plug, power, resistance, wire	blown, *current*, fuse, insulation, short circuit, three-core cable

Lesson plan

Reactivate knowledge

1 Two components are placed in parallel. How does their potential difference compare?

2 What happens the resistance of a circuit as more components are added in parallel?

3 What happens to the current from the supply when more components are added in parallel?

Answers: 1 They are the same *2* Decreases *3* Increases

Common misconception

Students may often think that a blown fuse explodes. Clarify that a fuse, at its simplest, is a very thin wire. This thin wire is designed to melt – not explode! – if the electric current flowing through it becomes too large. This could be because of an unexpected power surge, or a lightning strike in a storm. This is a safety feature in many household appliances and plugs.

Trigger interest

Demonstration: Short circuit (5 mins)

Connect the bulb and battery, showing it illuminated. Use the crocodile clips to carefully connect the thin wire in parallel with the bulb, shorting it (using this language early on). The wire will become very hot and could melt – students should observe this and describe what they see, giving ideas about what a short circuit means and why it may be hazardous inside a device (i.e., if in contact with flammable materials).

Emphasise the idea that more current will tend to flow down paths of the least resistance. Later, there will be opportunity to explore the implications of this in terms of the heating effect on thin wires.

Exposition

Plugs, cables, and short circuits (5 mins)

Summarise the purpose of each of the three wires in a plug. Some students may be familiar with this, so mini-whiteboards are useful here to test prior knowledge. The main activity can be carried out in tandem with a live demonstration of showing how a plug is unscrewed and talking through the wires therein. Ensure the plug is disconnected appropriately.

Provide students with coloured pens to annotate the image of the plug and encourage discussion about the choice of materials.

Be aware that students who may be colour blind or visually impaired will need tailored support – teachers may need to ask these students specifically how coloured wires appear to them, and use appropriate visual aids (e.g., matching coloured pens to what they see on diagrams). Ensure their understanding is checked on a one-to-one basis.

Main activity

Plugs, cables, and short circuits (40 mins)

Students should answer the questions in pairs. There is a substantial amount of information to digest here, so break up the lesson with mini-plenaries to ensure students can both recall key features of a three-core cable, and they can understand the choice of materials.

To extend students' understanding, explore the heating effect and higher resistance of thin wires. Ask students to suggest explanations for this: they should consider the fact that there are less electrons (per unit of area) in a thinner wire and so these electrons have to move even faster for the same current (i.e., the same amount of charge flowing per unit of time).

Review and practice

Wonky wiring (5 mins)

Students use the interactive to match the colour of a wire's insulation with which pin of a plug it is attached to. They then complete a paragraph to summarise the hazards associated with plugs and sockets.

If there is time, extend this activity by linking to a career case-study of an electrical technician, e.g., via academic or vocational study. Students could imagine and suggest the types of things they would have liked in physics lessons and what types of skills they would apply in their day-to-day jobs, e.g., analysing and problem solving, creativity, and technical knowledge (knowledge of electric circuits, current, resistance, etc).

P5.3 Electrical power and potential difference

AQA GCSE Physics specification points

4.2.4.1 Power

4.2.4.2 Energy transfers and everyday appliances

Smart Curriculum Narrative

Electricity and magnetism

Theme/subtheme

Electricity and magnetism/Domestic electricity

Disciplinary knowledge

- WS 1.2, 1.4, 4.5
- MS 3b, 3c

Activity

Applying the power equations

Learning objectives	Learning outcomes		
	Developing Aiming for 3	Secure Aiming for 5	Extending Aiming for 8
Calculate energy transferred by a device in a set time using its power rating	State that the power of a device is the amount of energy transferred by it each second	**Calculate energy transferred by a device in a set time using its power rating**	Rearrange power equation to find either power, energy or time and convert units
Calculate the power of electrical devices using potential difference and current	State the equation to calculate power	**Calculate the power of electrical devices using potential difference and current**	Rearrange the power equation and determine appropriate fuse value for a device
Calculate the power loss by heating using resistance and current	State that a current has a heating effect	**Calculate the power loss by heating using resistance and current**	Derive and use the equation for power using current and resistance

Tier 2 vocabulary	Tier 3 vocabulary
current, electric, energy, joules, transfer	efficiency, electrical power, fuse, mains, potential difference, power, power rating, supply, work done

Lesson plan

Reactivate knowledge

1 How are power and energy related?

2 Give the units of energy and power.

3 A student uses a 15 W then a 30 W motor to lift a small load a set height. What did they observe?

Common misconception

Students may think that the power rating on everyday appliances tells you how much power they produce. Tell students that the power rating on everyday appliances tells you the amount of energy per second they draw from the mains power supply. The power rating also indicates the maximum power at which the device can safely operate.

Answers: 1 Power is the energy transfer per second *2* Power is measured in watts (W) and energy in joules (J) *3* The 30 W motor would lift the load faster

Trigger interest

Equation gap-fill (10 mins)

List a set of incomplete equations with gaps for students to complete. Design this to build on students' prior knowledge, so that by process of elimination, they can determine the definition of electrical power (as the rate of energy transfer) independently. Support students by using word equations wherever possible.

Additional challenge can be incorporated using symbols and units. Encourage students to think about dual applications: for instance, the letter 'W' could be matched with the symbol for 'work done', or it could be the unit of power, matching with 'joules per second'. Support students who may find this tricky by referring to word equations.

This is mathematical lesson, centred on applying equations to various contexts – therefore, explore ways to gamify the key learning where appropriate, reinforcing the defining equations for electric current, potential difference (p.d.), and power.

Exposition

Applying the power equations (15 mins)

Give students the opportunity to practise some simple applications of the power equations they were introduced to at the start of the lesson. As usual, at least a couple of examples should be given to embed them, ideally including units of kW and/or mA.

Show how the equations $V = IR$ and $P = VI$ can be combined algebraically, to get the equation for the heating effect of a current, $P = I^2R$.

Students should suggest why this equation is called 'the heating effect of current' before attempting some simple calculations.

Main activity

Energy and power (25 mins)

In this practical, recap the concept that energy is transferred between different stores, and the importance of considering efficiency for electrical appliances in the home. Ask students what factors may affect the amount of energy delivered by a current, with reference to the equation $P = VI$.

Use the practical to explore the amount of energy transfer over one minute for a variety of components.

Review and practice

Match the fuse (5 mins)

Students need to select the most appropriate fuse for an electrical appliance after being told the power rating. This involves calculating the current (assuming the appliance is connected to a mains p.d. of 230 V) and then choosing the fuse that is slightly higher. Use 3 A, 5 A, 13 A, and 30 A fuses. Ensure students can justify why a lower-rated or higher-rated fuse would be inappropriate.

Students may connect well with this activity if, wherever possible, appliances are selected that match their interests. Scope these with the students, by finding out their most used electrical item. This activity can be expanded by asking students to independently research the power ratings of these appliances before selecting appropriate fuses.

P5.4 Electrical currents and energy transfer

AQA GCSE Physics specification points

4.2.4.1 Power

4.2.4.2 Energy transfers in everyday appliances

Smart Curriculum Narrative

Electricity and magnetism

Theme/subtheme

Electricity and magnetism/Domestic electricity

Disciplinary knowledge

- WS 1.2, 1.4, 4.5
- MS 3b, 3c

Practical

The power of lamps

Learning objectives	Learning outcomes: Developing Aiming for 3	Learning outcomes: Secure Aiming for 5	Learning outcomes: Extending Aiming for 8
Describe how work is done when charge moves through a circuit	State that electrical appliances are designed to transfer energy	**Describe how work is done when charge moves through a circuit**	Explain how energy is conserved in a circuit
Calculate the energy transferred by electrical work	State the equation used to calculate energy transferred in a circuit	**Calculate the energy transferred by electrical work**	Perform calculations involving rearrangement of the equation $E = VQ$
Describe the changes in energy stores of different appliances	Identify the factors that affect the energy transfer in a circuit	**Describe the changes in energy stores of different appliances**	Use algebra to combine the equations $Q = It$ and $E = VQ$ to form the relationships $E = VIt$ and $P = IV$

Tier 2 vocabulary	Tier 3 vocabulary
conserved, dissipated, electron, energy, flowing, thermal energy, transfers	ammeter, charge, component, *current*, lamp, parallel, potential difference, variable resistor, *series*, useful, voltmeter

Lesson plan

Reactivate knowledge

1 Describe the changes to how energy is stored when a battery-operated fan is switched on.

2 How do you calculate how much charge has flowed in a set time?

3 What is potential difference?

Common misconception

Students may think thermal energy is lost in a lamp. Clarify that energy is never *lost*; it is always conserved. The correct terminology in this case is to say that charge *does work on* a lamp as it flows through. *Does work on* is equivalent to saying *transfers energy* to. Therefore, charge transfers energy to the thermal store of a lamp which is dissipated and wasted as it is not useful here.

Answers: 1 The chemical store decreases as energy transferred by the current to the kinetic store of the motor in the fan. *2* Charge = current × time *3* Energy transferred per unit charge

Trigger interest

Energy gain or drop? (5 mins)

Frame the start of this lesson as story about an electron's journey around a circuit as part of a d.c. current. The idea is to gradually build this concept to think about units of charge flowing as current.

Draw a simple series and a simple parallel circuit, each containing two lamps and a cell. Above each component, ask students to select between two words, 'gain', or 'loss', to describe how the energy carried by the electron changes as it flows through each component.

Develop this concept by asking students to think about units of charge (i.e., many electrons) and linking this to potential difference (i.e., the energy transferred per unit of charge across each component). Extend students' understanding by further examining the use of the word 'loss', in reference to energy transfer, and the misconception this might infer.

Exposition

Charge and current (10 mins)

Show students a simulation or model of a direct electric current and discuss the movement of electrons around the circuit. As in the starter activity, show both series and parallel examples, emphasising the 'choice' of route at a junction in parallel circuits. Ensure that students understand that charge is conserved around the circuit.

Students should be able to recall the equation $Q = It$ or $I = \frac{Q}{t}$.

It is useful at this stage to practise a few calculations with it, checking their values with a virtual ammeter.

With the terminology of energy transfers being 'gains' or 'drops', use the idea of a potential difference being the way to quantify how much energy transfer (per unit of charge) takes place across a component. Provide a couple of simple examples using $V = \frac{E}{Q}$.

Main activity

The power of lamps (30 mins)

Students work through the practical using the variable resistor to gradually increase the current through a lamp, measuring the current and potential difference across it using an ammeter and a voltmeter, and describing changes to its brightness.

Review and practice

Electric crossword (10 mins)

Students use the interactive to complete the crossword. They will then select keywords to complete a summary paragraph that reinforces the key concepts of this lesson.

Extend students by giving them the completed crossword and asking them to write the clues.

P5.5 Appliances and efficiency

AQA GCSE Physics specification points

4.2.4.2 Energy transfers in everyday appliances

Smart Curriculum Narrative

Electricity and magnetism

Theme/subtheme

Electricity and magnetism/Domestic electricity

Disciplinary knowledge

- WS 1.2, 1.4, 4.5
- MS 3b, 3c

Activity

Electrical energy transfer

	Learning outcomes		
Learning objectives	**Developing Aiming for 3**	**Secure Aiming for 5**	**Extending Aiming for 8**
Calculate the energy transferred by an appliance using the power rating and time it is on for	Describe the factors that affect the cost of using various electrical devices	**Calculate the energy transferred by an appliance using the power rating and time it is on for**	Analyse the use of a variety of electrical devices to determine their cost of operation
Calculate the useful output power of an appliance using efficiency equation	Describe how electrical appliances waste energy	**Calculate the useful output power of an appliance using efficiency equation**	Compare a range of electrical devices in terms of efficiency using calculations to support any conclusions
Calculate energy transferred to an appliance using the current, potential difference, and time	State that electrical power is the product of the potential difference and the current	**Calculate energy transferred to an appliance using the current, potential difference, and time**	Use and rearrange the energy transferred equations in a wide variety of calculations

Tier 2 vocabulary	Tier 3 vocabulary
consumption, *current*, electrical, energy, frictional force, heating, power, transfer	efficiency, efficiency rating, mains, inefficiency, potential difference, power rating

Lesson plan

Reactivate knowledge

1 How does the power of a device relate to potential difference across it and current through it?

2 How can you calculate the energy transferred by a device using its power rating and time it was on for?

3 What is the efficiency of a device?

Common misconceptions

Students may think that the power rating on everyday appliances tells you how much power they produce. Clarify that the power rating on everyday appliances tells you the amount of energy per second they draw from the mains power supply. The power rating also indicates the maximum power at which the device can safely operate.

Answers: 1 Power = potential difference × current *2* Since power is rate of transfer of energy, Energy = power × time *3* The proportion of total input energy/power usefully transferred or (useful energy transferred/total energy input)

Trigger interest

Efficiency rating (10 mins)

Discuss the wider context around UK law and the requirement to label all appliances with an efficiency rating/sticker. Show a few examples of these; in advance of the lesson, students could find examples of these at home and send pictures to a shared slide-deck.

Discuss the rationale for this requirement and the advantages for consumers to be energy literate, especially students who can calculate things like electrical power consumption and efficiency, to properly manage their personal finances.

Exposition

Powerful recap (10 mins)

Reintroduce the definition of power in words, and ask students to write this as an equation, using correct symbols, on mini-whiteboards.

Ask them to rearrange this equation so that energy is the subject (students who are able to do this without support can go straight to the main activity).

Provide additional support to rearranging $P = \frac{E}{t}$ to $E = Pt$ as required.

Equation triangles might be useful here. Ensure students can substitute $P = VI$ in place of power P, in both equations, and emphasise why this will be useful (because one can calculate the amount of energy transferred by measuring the current and potential difference directly).

Main activity

Electrical energy transfer (25 mins)

Remind students that they have met the concept of efficiency in P1.9. With reference to some typical household appliances, remind students that some energy transfers are useful and others are not, and how this can be used to calculate efficiency. Hence, appliances that do the same thing (e.g., light bulbs) can be compared by their efficiency.

Students should also discuss the reasons for inefficiency; the heating effects of a current and of frictional forces.

Review and practice

Comparing appliances (10 mins)

Choose an appliance, e.g., a toaster or a microwave, and provide students with an image of a UK and US model, (ideally showing their connecting plugs) Give students data about them, including the mains p.d. they are connected to, the current supplied, and operating time. Students should identify which appliance is the most efficient.

Extend students' understanding by asking them to identify which appliance is manufactured for use in the US and UK, expecting them to justify their answers by referencing the different plug connections. They can study some facts about UK plugs by revisiting P5.2.

P5 Electricity in the home

Introduction to checkpoint intervention

The checkpoint intervention provides suggestions for a lesson to follow up the *P5 Electricity in the home* checkpoint assessment. Depending on the outcomes of the assessment, these suggestions can be used to move a class towards secure, to target specific learning outcomes, or to consolidate knowledge for students achieving secure outcomes. Use students' outcomes from the checkpoint assessment to address general misconceptions from the content covered. Be prepared to re-cover content as required; three key concepts from the chapter are covered in more detail below.

Checkpoint secure learning outcomes

- Describe the difference between direct and alternating potential difference.
- Describe why a short circuit inside a device presents a hazard.
- Describe how work is done when charge moves through a circuit.
- Calculate the useful output power of an appliance using the efficiency equation.
- Explain why the National Grid is an efficient system to transfer energy.
- Calculate the power of electrical devices using potential difference and current.

Student reflection

Encourage students to reflect on whether there were any checkpoint questions they found difficult or straightforward as well as their preparation for the checkpoint assessment, including how they revised, what they revised, the time spent on revision, and what areas they could improve upon. For more strategies, see the Metacognition spread.

Intervention activity: Describe the difference between direct and alternating potential difference.

Recognising alternating current and direct current

The circuit symbol for an alternating power supply is a useful starting point. Ask students to describe what they see and build on the fact that the circuit symbol is a useful clue to what an alternating voltage or current actually looks like. Waves oscillate, or go 'to-and-fro' in cycles, just as charges in a circuit do, when connected to an alternating voltage. The contrast to this motion and direct motion should be emphasised.

Simulations and roleplay

Any simulations that clearly show electron drift (in one direction) under a direct voltage, and oscillation under an alternating voltage, will work effectively. Role play, using the learning space wherever possible, can also get students to think about expressing the difference between direct and alternating current kinaesthetically. For students who struggle to describe the motion verbally, provide a simple gap-fill activity with keywords and phrases, such as oscillate, or one-direction. Finally, a sketch of potential distance against time for direct current and alternating current should be compared and discussed, linking this back to the alternating current supply circuit symbol. It is worth emphasising the fact that the p.d. for an alternating current supply varies continuously between positive and negative. This is not the case for direct current (it can, for instance, vary positively). Expect students to quickly identify that the sinusoidal curve corresponds to alternating current and ensure they can see the potential difference switches direction (goes from positive to negative) in one complete cycle.

Intervention activity: Describe how work is done when charge moves through a circuit.

Work done means energy transfer

Students tend to be comfortable describing energy transfers directly – confusion may arise when introducing concepts of work being done. It is important to emphasise the equivalence between the phrases *energy transferred to* and *work is done on*. Another way to gently illustrate this equivalence is to ask students what the unit of energy and work is, and what this means about the two quantities.

Narrating charge's journey through a circuit

Students could factually describe the journey of a unit of charge as it completes a loop around a circuit containing two lamps, or, if appropriate, they could turn this factual account into a story of the electron's journey from a first-person perspective.

Provide a circuit diagram, and a pre-made circuit where possible (series at first, parallel if appropriate). Ensure students begin their story at the cell. It might start as, "Once the circuit is complete, and the lamps immediately light up, the unit of charge 'wakes up' and collects energy from the cell, beginning its journey ...". Guide students to quantify this collection of energy, for example, a 6 V cell means 6 joules of energy per coulomb. Students can develop their stories by including phrases about doing work on components, and ideas about conserved charge and energy (i.e., they must return to the cell with exactly zero joules per coulomb!) and how many journeys can be made considering the finite lifetime of the cell.

Intervention activity: Calculate the power of electrical devices using potential difference and current.

Identifying and interpreting appliance information

Students can improve their confidence by practising looking for the electrical power and efficiency information on a variety of appliances. They can do this independently at home, collecting images to discuss and analyse together in pairs or groups. They should become confident at recognising the operating power, current, and mains potential difference, before attempting to apply the equations.

Equation practice

Be aware that units on everyday appliances may appear unfamiliar to students at first glance.

Ensure they are able to distinguish units of amps, volts, and watts from unit prefixes as an intermediate activity. Finally, once they are ready to apply the equations to calculate power directly, use equation triangles wherever needed to give students a framework for their substitutions and calculations, especially if rearranging. Ensure that straightforward examples are used to build confidence and familiarity, before progressing to contexts that use units such as kW, mA, and so on.

P6 Molecules and matter

P6.1 Density

AQA GCSE Physics specification points

4.3.1.1 Density of materials

***Smart Curriculum* Narrative**

Forces and motion

Theme/subtheme

Forces and motion/Pressure

Disciplinary knowledge

- MS 1a, 1b, 1c, 3b, 3c, 5c

Required Practical 5

Density

	Learning outcomes		
Learning objectives	**Developing Aiming for 3**	**Secure Aiming for 5**	**Extending Aiming for 8**
Explain what density means	Identify materials with high and low densities	**Explain what density means**	Explain the difference between density and mass
Calculate the density of materials	State that the density of a material is the mass per unit volume	**Calculate the density of materials**	Use and rearrange the density equation in a wide variety of calculations
Determine the density of a range of solids and liquids	Determine the volume of some regular shapes and irregular shapes with support	**Determine the density of a range of solids and liquids**	Evaluate in detail the experimental measurement of density, considering resolution of instruments and accounting for errors in measurements

Tier 2 vocabulary	Tier 3 vocabulary
cubic metres, *balance*, *float*, irregular solid, joule, kilogram, liquid, measuring cylinder, meniscus, range, regular solid, *resolution*	density, mass, unit volume

Lesson plan

Reactivate knowledge

1 What are the units of mass?

2 What is mass?

3 How do you find the volume of a cube?

Common misconception

Students may think that cutting an object in half, halves the density of each piece is half that of the original piece. Clarify that density is a ratio of mass to volume. If the mass halves alongside the volume, the density of the object stays the same. It has no relation to the number of pieces that an object is divided into, nor to the size of the object.

Answers: 1 g/kg *2* The amount of matter in something *3* Length × width × height

Trigger interest

Demonstration: Comparing cans of fizz (10 mins)

Fill a small tank of water – enough to submerge a can of fizzy drink. Ask students to predict which cans will float and which will sink. First, compare 'regular' and 'diet' flavours, ensuring both cans are the same size and volume. Then compare two cans of different sizes but with the same contents.

Encourage students to make predictions and identify the assumptions that they are making. This is a great demonstration to reveal prior knowledge and/or any misconceptions the students have about density. Expect that students are aware that diet versions have less sugar and hence are less dense than their regular counterparts.

Exposition

How heavy? (10 mins)

Explore the common misconception about weight and an object's propensity to sink or float. Images of large floating objects like ships, contrasted with small and dense sunken objects are good to discuss. Bear in mind that misconceptions about density are often very difficult to change, even in light of observational evidence! Emphasise that when considering if something sinks or floats, the way in which an object's mass is distributed in space is just as important as how much there is.

Introduce the definition of density as a ratio between mass and volume – first in words and then as an equation. Expect students to be able to state units and draw attention to the Greek symbol rho (ρ). In preparation for the practical, use the Maths Skills interactive to provide students with some mass and volume data to calculate the density some materials using the equation.

Support students by providing the equations for volumes and some example calculations. Where appropriate, extend students by discussing units of kg/m^3 and g/cm^3.

Main activity

Density (30 mins)

Students carry out the practical to measure the density of some samples. They should focus on the resolution of the instruments and the appropriate use of significant figures in their answers.

Challenge some students to consider the percentage of uncertainty in their overall density answers, given the resolution of their measuring equipment. For instance, they could determine the percentage uncertainty in the mass and volume readings, and then add these to work out the overall percentage uncertainty in the value obtained for density.

Review and practice

Smoke signals (5 mins)

Give pairs of students a wooden splint and light them. Students should blow it out after a few seconds, and discuss what they observe.

Scaffold their responses with prompts about the rising smoke – linking this to the idea of gasses floating on top of each other due to density differences.

P6.2 States of matter

AQA GCSE Physics specification points

4.3.1.1 Density of materials

Smart Curriculum Narrative

Forces and motion

Theme/subtheme

Atomic and nuclear physics/Particle model and changes of state

Disciplinary knowledge

- WS 1.2

Activity

Changing state

	Learning outcomes		
Learning objectives	**Developing Aiming for 3**	**Secure Aiming for 5**	**Extending Aiming for 8**
Use the particle model to explain the difference in density of solid, liquid, and gases	Draw a diagram to show the arrangement of the particles in a solid, liquid, and gas	**Use the particle model to explain the difference in density of solid, liquid, and gases**	Explain differences in density between two materials in the same state
Describe the changes in behaviour of the particles in a material during changes of state	Name the changes of state	**Describe the changes in behaviour of the particles in a material during changes of state**	Describe the changes in the energy of individual particles during changes of state
Describe how mass is conserved during a change of state	State that a change of state is a physical change because the material recovers its original properties when the change is reversed	**Describe how mass is conserved during a change of state**	Explain the difference in physical and chemical changes using ideas about particles

Tier 2 vocabulary	Tier 3 vocabulary
density, energy, ice, mass, matter, particle, physical change, random, steam, vibrate, *volume*, water vapour	boil, condense, conservation of mass, freeze, kinetic theory, melt, sublimation, vaporisation

Lesson plan

Reactivate knowledge

1 In which states of matter do the particles move randomly?

2 In which state of matter are the particles furthest apart?

3 What happens to the particles in a substance as you heat it?

Common misconception

Students may think that particles in a solid are fixed in position and have no energy. Clarify that particles in a solid vibrate around a fixed position, therefore, they do have energy. However, relative to their configuration in a liquid or gas, they have less energy.

Answers: 1 Gases and liquids *2* Gas *3* The particles gain more energy/they move vibrate faster

Trigger interest

Demonstration: Changing state (10 mins)

Model the particles of the three states of matter using small balls and transparent containers. Ideally, use three students to assist, each with their own container.

Place a few balls into a container and shake them around. This is similar to a gas, because the balls move around freely and are generally far apart. Then, almost fill a container – the particles can still move a bit, but there are a few gaps between them. This is similar to a liquid. Finally, fill a container so the balls cannot move – they are closely packed together in a manner similar to a solid. Throughout this demonstration, emphasise that changing the number of balls in a container equates to a 'change in state', and describe the changes in the forces and relative energy of the particles between state changes.

Extend students by asking them to consider the inaccuracies of the model in demonstrating the differences in particle motion in different states, and the potential misconceptions that may arise. For instance, in a liquid, particles can 'slip' and move around each other – this isn't easy to demonstrate in this model, nor is the constant vibration of particles about a fixed position in the solid state.

Exposition

Recapping states of matter (10 mins)

Place a beaker of ice melting in some warm water on a set of scales. Ask a student to keep an eye on this – they are to record the initial and final mass at the beginning and end of the lesson – and report back to the class. Draw out ideas about why this observation is important, and lead students to conclude that mass is conserved during a change of state.

Use the states of matter PhET simulation to demonstrate key properties of solids, liquids, and gases. If possible, allow students to explore this independently and circulate the room to provide support/challenge alongside this interactive tool as needed.

Additional support can be given by providing a list of keywords and phrases to help students to verbalise their observations (e.g., vibrating, moving freely, and random).

Main activity

Changing state (25 mins)

Students use ideas from the simulation to complete the worksheet. Diagrams can be made more interactive by using scrap paper discs from a hole-punch to represent particles of matter and the forces between them.

Students could develop ideas further by using springs and small balls of sticky tack or soft modelling clay to represent solid lattices. In this case, the small balls represent the particles of matter, and the springs represent the bonds between the particles. These could be 'broken' by input of energy, representing a change of state (i.e., from solid to liquid).

Review and practice

Property match (10 mins)

Interactive where students sort the properties according to whether they describe solids, liquids, and gases. They then match descriptions of other properties – such as density, fluidity, and compressibility – with solids, liquids, and gases.

If there is time, extend this activity by referencing plasma as the fourth state of matter. A plasma ball is a nice demo to accompany this activity. Some students may already know what a plasma is, so draw out some knowledge and make links to some surprising ways in which plasmas are seen in everyday life, e.g., lightning strikes and neon lights. Highlight that plasma physics is a rich area of research with many cutting-edge implications, like understanding how the sun's plasma ejections affects satellites orbiting the Earth.

P6.3 Changes of state

AQA GCSE Physics specification points

4.3.1.2 Changes of state

4.3.2.3 Changes of state and specific latent heat

Smart Curriculum Narrative

Energy

Theme/subtheme

Energy/Internal energy

Disciplinary knowledge

- WS 3.5
- MS 4c

Practical

Measuring the melting point of a substance

Learning objectives	Learning outcomes: Developing Aiming for 3	Secure Aiming for 5	Extending Aiming for 8
Identify melting or boiling points on a heating or cooling graph	State the definitions of melting and boiling points	**Identify melting or boiling points on a heating or cooling graph**	Evaluate data produced by a heating experiment to discuss the reproducibility of the measurement of a melting point
Describe how the melting points and boiling points of a substance can be changed	State that the melting and boiling points of a pure substance are fixed	**Describe how the melting points and boiling points of a substance can be changed**	Explain how adding impurities changes the melting point of a substance
Use the term 'latent heat' to describe the energy gained by a substance during heating for which there is no change in temperature	Describe the process of melting and boiling	**Use the term 'latent heat' to describe the energy gained by a substance during heating for which there is no change in temperature**	Explain why there is no temperature change during a change of state

Tier 2 vocabulary	Tier 3 vocabulary
atmospheric pressure, energy, particle, temperature, thermometer, vapour	latent heat, temperature–time graph

Lesson plan

Reactivate knowledge

1 Name the change of state when a solid becomes a liquid.

2 What is the state of 45° wax if it has a melting point of 50°?

3 What is the state of 80° ethanol if it has a 78° boiling point?

Common misconception

Students may think evaporation is the same as boiling. Emphasise that evaporation can happen at any temperature, whereas boiling only happens at the substance's boiling point. This can be easily understood by considering a puddle on a road on a sunny winter's day: it evaporates, it doesn't boil!

Answers: 1 Melting *2* Solid *3* Gas

Trigger interest

How hot? (5 mins)

Provide students with some important temperatures and ask them to match them to images of objects or changes (e.g., boiling and freezing point of water, temperature of the surface of the Sun, average human body temperature, etc).

Extend students by providing more obscure temperatures to discuss, for example, –273 °C being absolute zero, or by including temperatures in different units (e.g., Fahrenheit).

Exposition

Does temperature always increase? (5 mins)

Recap the changes in behaviour of the particles as a solid melts and then boils. A video or simulation (PhET, Phase Changes), clearly showing the increase in vibrational motion, bonds breaking, and particles becoming free to move and slide across one another, works well here. Emphasise that energy is always being delivered to the system throughout the process; the idea is to then set up the hinge question that frames the practical activity – during a change of state, does the energy supplied increase the temperature of the object? Or, is the energy used to do something else?

Have a short paragraph displayed that summarises the changes in particle behaviour as a material melts, to refer to during the practical.

Main activity

Measuring the melting point of a substance (35 mins)

The students carry out the practical and produce a temperature–time graph. The region in which the state changes (no increase in temperature) should be clearly identified. Lead students to their own conclusions – scaffolding questions as appropriate. They should appreciate that temperature remains constant, even while energy is being added to the material, and that this happens during the change of state. This is a surprising result! Try to get students to appreciate that changing state is a strange phenomenon. Ensure ideas about substance purity are discussed – that adding impurities alters the melting and boiling point of a substance (e.g., salt water).

A useful contrast to employ is an incorrect temperature–time graph (e.g., a straight line with positive gradient). Students should be able to understand that this would mean the temperature always rises, which is probably what most might expect! However, this incorrect graph implies no change of state takes place.

Revisit the hinge question with individuals or small groups, and ensure they conclude that temperature doesn't change during a change of state, and that they can identify and label the melting and boiling regions on their graphs. The increasing energy store is not obvious, hence the term, 'latent heat'. This is a useful opportunity to define the meaning of 'latent'.

Review and practice

Boiling at altitude (10 mins)

Show students a graph of the boiling point of water compared with altitude. Students use the interactive to complete a description of this relationship and an explanation of why the boiling point changes.

P6.4 Internal energy

AQA GCSE Physics specification points

4.3.2.1 Internal energy

4.3.2.2 Temperature changes in a system and specific heat capacity

Smart Curriculum Narrative

Energy

Theme/subtheme

Energy/Internal energy

Disciplinary knowledge

- MS 1a, 3b, 3c, 3d

Demonstration

Internal energy

Learning objectives	Learning outcomes: Developing Aiming for 3	Learning outcomes: Secure Aiming for 5	Learning outcomes: Extending Aiming for 8
Describe how the internal energy is the sum of the all the kinetic and potential energies of the particles in the system	State that internal energy is the energy stored inside a system by its particles	**Describe how the internal energy is the sum of the all the kinetic and potential energies of the particles in the system**	Use the concepts of kinetic and potential energy to explain changes in internal energy
Describe how heating causes either a change in temperature or a change in state	State that internal energy increases as a system is heated	**Describe how heating causes either a change in temperature or a change in state**	Explain how kinetic and potential energy of the particles changes at a change of state
Describe how the behaviour of particles changes as the energy of a system increases	Outline the behaviour of particles in solids, liquids, and gases	**Describe how the behaviour of particles changes as the energy of a system increases**	Describe the changes in the size of intermolecular forces during changes of state

Tier 2 vocabulary	Tier 3 vocabulary
boil, energy, force of attraction, gas, heating, liquid, mass, molecule, particle, random, solid, steam, temperature, vapour	change in state, energy transfer, internal energy, kinetic energy, potential energy, specific heat capacity, specific latent heat

Lesson plan

Reactivate knowledge

1 What happens to the temperature of a substance as it melts?

2 What determines how much the temperature of a substance rises as it is heated?

3 What is the name given to the energy required to change state?

Common misconception

Students may think when a substance is heated, the particles expand or get bigger. Clarify the particles gain more energy it moves around more and takes up more space.

Answers: 1 It stays the same *2* Energy transferred, mass, and specific heat capacity *3* Latent heat

Trigger interest

Specific heat capacity (10 mins)

Recap the equation $\Delta E = mc\Delta\theta$, then ask the students complete the interactive where they work through some example calculations.

Emphasise that this equation is linked to changes in temperature – ask students to debate whether this equation applies to changes in state, considering the effects they observed in the previous lesson. They should conclude that it does not apply – but that there still is addition of energy, recalling the idea of latent heat.

Exposition

What is internal energy? (5 mins)

Using a simple image of gas molecules in a box, or a simulation showing them in motion, introduce the concept of dual energy stores of the particles being their kinetic and potential energies. The internal energy of the particles is the sum of these stores – the fact that 'internal' does not include gravitational potential energy, or the kinetic energy caused by motion of the whole substance, should be emphasised.

Main activity

Demonstration: Internal energy (30 mins)

Set up a beaker of water and crushed ice, with a thermometer and data logger (if available). The temperature–time graph can be plotted in real-time, as the students complete the worksheet.

Begin heating the beaker of iced water and discuss the energy transfers involved. Initially, there will be no temperature increase, and students should recall from the previous lesson that this is related to latent heat. Scaffold the discussion so they can conclude that the latent heat only increases the potential energy of the particles during a change of state – at this point, discuss the repositioning and weakening of the bonds between the particles. The PhET simulation, Phase Changes, again works well here and can be run alongside.

As the ice melts and the temperature starts to increase, lead students to describe the changes in motion of the particles, and the changes to the internal energy, in terms of changes to kinetic and potential energy. Note that evaporation occurs constantly through this process; when the water is boiling, ask students to observe the temperature of the boiling point. With the demonstration and completed worksheet, they should have a good understanding of internal energy.

Extend students by asking them to reflect on the intermolecular forces of attraction as system energy changes. Where appropriate, link to their prior knowledge of Van der Waals forces.

Review and practice

True or false? (10 mins)

Quiz students on a variety of true or false statements summarising key ideas from changes of state and internal energy. Use this opportunity to differentiate questions appropriately and explore some of the common misconceptions around boiling and evaporating, changes of energy, and key terms.

P6.5 Specific latent heat

AQA GCSE Physics specification points

4.3.2.3 Changes of state and specific latent heat

Smart Curriculum Narrative

Energy

Theme/subtheme

Energy/Energy and temperature

Disciplinary knowledge

- MS 1a, 3b, 3c, 3d, 4a

Practical

Specific latent heat

Learning objectives	Learning outcomes: Developing Aiming for 3	Learning outcomes: Secure Aiming for 5	Learning outcomes: Extending Aiming for 8
Describe specific latent heat in terms of energy changes that occur in a change of state	State the energy changes that occur at a change of state	**Describe specific latent heat in terms of energy changes that occur in a change of state**	Explain the difference between specific heat capacity and specific latent heat
Calculate the energy required for a change of state using data	Describe factors affecting energy required for a change of state	**Calculate the energy required for a change of state using data**	Complete calculations where there is a change in temperature and a change in state
Describe the difference between specific heat capacity and specific latent heat	State that both specific heat capacity and specific latent heat are properties of a substance	**Describe the difference between specific heat capacity and specific latent heat**	Explain why scalding by steam is much more dangerous than scalding from boiling water

Tier 2 vocabulary	Tier 3 vocabulary
balance, *bond*, condense, degree Celsius, energy, freeze, ice, immersion heater, joule, joulemeter, kilogram, low-voltage heater, mass, melt, particle, watt	energy transfer, fusion, latent heat, specific latent heat, specific latent heat of fusion, specific latent heat of vaporisation

Lesson plan

Reactivate knowledge

1 In which state does a substance have the most internal energy?

2 What is the specific heat capacity of a substance?

3 Is energy transferred to or away from a substance as it condenses?

Common misconception

Students may confuse specific latent heat with specific heat capacity. Clarify that specific latent heat refers to the energy supplied/released when intermolecular bonds are broken/formed. The temperature of the object stays constant during this process, which is why it is called 'latent' (or hidden) heat. Specific heat capacity refers to the amount of energy required to raise an object's temperature (specifically 1 kg by 1 °C).

Answers: 1 Gas *2* The energy transferred to change the temperature of 1 kg of the substance by 1 °C *3* Away from

Trigger interest

Puddle puzzle (10 mins)

Introduce the scenario of a student learning about changes of state, but being puzzled by an outdoor puddle that seems to turn into a gas well below its boiling point. Concept-map some ideas from students about why this happens – expect some recall from previous lessons about the difference between evaporation and boiling, and other external factors affecting the rate of evaporation like wind and surface area.

Students then use the short interactive to rearrange sentences to form an explanation of why puddles evaporate.

Exposition

Chunking the theory (10 mins)

This lesson contains a lot of new terminology, equations, and synthesis of knowledge from previous lessons, so chunk activities and exposition as appropriate. Break the theory into two parts.

Part one – specific latent heat of fusion: Show a simple image of ice melting, ideally with a heat source. Assuming an initial temperature of, say, −3°C, and a final temperature of 4°C, sketch four temperature–time graphs, including incorrect options that show no latent heat/change in state (i.e., no flat line). Students should discuss and vote for the correct one on mini-whiteboards, justifying their responses.

Recap the energy changes during the heating and melting process, with an emphasis on the two stages: rising temperature and changing state. Add labels to sketches of graphs wherever possible.

Emphasise the weakening of bonds during melting, and note that the bonds are yet to be fully broken (alluding to the upcoming boiling process). Introduce the terminology of specific latent heat of fusion, discussing why fusion is an appropriate term to describe this process. Students then complete the Maths Skills interactive to practise using the equation.

Part two – specific latent heat of vaporisation: Show a simple image of water boiling, ideally with a heat source, and again give voting options for the correctly sketched temperature–time graph. Students should identify the flat section of the graph as the change of state and the sloped line as changing temperature. Introduce the definition of specific latent heat of vaporisation and discuss why vaporisation is an appropriate term to describe this process. Check with a final simple calculation that students can apply the equation before moving to the practical. Challenge students to state and explain which would be larger for the same mass of water – the specific latent heat of fusion or vaporisation.

Main activity

Specific latent heat (30 mins)

Form students into groups and get them to randomly select one of the two practical sheets. They should share data with another group doing the same experiment to find a mean value for latent heat. This is an opportunity to develop their collaborative skills. Ensure inclusive participation.

Review and practice

Overall heating (5 mins)

Return to the image of ice melting used at the start of the lesson. Ask students to perform a simple calculation that involves both equations learned, giving initial and final temperature and asking for the final energy change. This question can be differentiated appropriately. For example, support students by providing a calculation template to simplify the problem for some students. They may need prompting that they will need two separate equations. Additional challenge can be provided by omitting the fact that the problem requires both equations, or by giving red-herring data.

P6.6 Gas pressure and temperature

AQA GCSE Physics specification points

4.3.3.1 Particle motion in gases

Smart Curriculum Narrative

Forces and motion

Theme/subtheme

Forces and motion/Pressure

Disciplinary knowledge

- W.S 1.2, 1.3, 1.4
- MS 4b

Practical

Gas pressure and temperature

	Learning outcomes		
Learning objectives	**Developing Aiming for 3**	**Secure Aiming for 5**	**Extending Aiming for 8**
Explain how a gas can exert a pressure	State that gas particles are in constant random motion	**Explain how a gas can exert a pressure**	Explain how the motion of the particles affects the gas pressure
Describe the relationship between an increase in the temperature of a fixed volume of a gas and the increase in pressure of the gas	State that increasing the temperature increases the pressure for a fixed volume of gas	**Describe the relationship between an increase in the temperature of a fixed volume of a gas and the increase in pressure of the gas**	Explain how the linear relationship between temperature and pressure for a fixed volume of gas can be used to determine absolute zero
Describe the behaviour of particles in a gas as the gas is heated	State that the temperature of a gas is related to the kinetic energy of the gas particles	**Describe the behaviour of particles in a gas as the gas is heated**	Explain why gas pressure increases with increasing temperature

Tier 2 vocabulary	Tier 3 vocabulary
density, gas, internal energy, liquid, microscope, molecule, particle, pascal, pressure gauge, random, solid, thermometer, *valve*	energy transfer, fixed volume, force, kinetic energy, pressure–temperature graph, unit area

Lesson plan

Reactivate knowledge

1 How does the pressure of a gas change if the gas is heated?

2 How do you calculate pressure?

3 How can you increase the pressure acting on a surface?

Common misconception

Students may think gases have no weight (or mass). Emphasise that mass is a measure of the amount of matter. In each state of matter the particles of matter don't vanish, they just have different amounts of internal energy. Imagine a drop of water is placed in a sealed plastic container (with all air excluded). It is then warmed until the droplet turns into water vapour. The total mass stays the same, because the number of particles inside the container stays the same.

Answers: 1 Gas pressure increases *2* Pressure is the force per unit area = force/area *3* Increase the force or decrease the area force acts on

Trigger interest

Demonstration: Theory of Brownian motion (15 mins)

Show Brownian motion using a smoke cell and discuss the conclusions that can be made using the Working scientifically sheet.

A single smoke cell can be connected to a visualiser so that the whole class can observe it – otherwise, show a video of the observed motion while individuals observe it through the microscope. Compare the observed motion of the smoke particles with pollen grains suspended in water, leading students to conclude that 'invisible' particles suspended in a fluid (i.e., a gas or a liquid!) must exist, because we can observe their effects directly. Emphasise that this motion was shocking and intriguing when scientists first observed it, because in the early 19th century, the idea of 'atoms' was still a matter of debate.

Exposition

Heating gas particles (10 mins)

Discuss the behaviour of gases in terms of particle motion and what happens to particles as their internal energy increases through heating. Ideally, use a gases simulation to demonstrate this in real time, scaffolding questions for students as the gas particles become more energetic.

Expect students to acknowledge collisions with the container walls, and link this to the idea of gas pressure. A balloon is a nice prop to use at this stage – ask for predictions for will happen to the gas pressure if the balloon is heated. Students should explain this in terms of the motion of particles: increased kinetic energy means more collisions (per second) with the balloon walls, until the balloon eventually bursts. Warn any students with sensory differences that there may be a sudden, loud noise. Simulations may also show containers 'bursting' open given enough energy through heating.

Students should be comfortable linking increased temperature to an increased (average) kinetic energy. Extend students further by asking them to consider why the potential energy of the particles is not considered for gases.

Main activity

Demonstration: Gas pressure and temperature (20 mins)

Frame this demonstration as testing the idea that increased temperature means an increased pressure. This gently introduces the students to the idea of proportionality – extend students by providing sample data so that they can plot a graph and describe this proportionality mathematically. They can explore ideas of absolute zero by extrapolating the temperature back to –273 °C.

Demonstrate the heating of a gas as outlined in the practical sheet.

Review and practice

The experimental method and gas laws (10 mins)

Emphasise that Boyle and Charles' experimental work on gases were important in leading to industrial advancements – i.e., understanding the relationship between pressure, volume, and temperature played a crucial role in the development of the steam engine. Students should appreciate that physics plays a crucial role in advancing industry and technology innovations, which are important to any society's economy, health, and prosperity.

This is also an opportunity to touch on the global development and ongoing exchange of scientific knowledge, emphasising that historical scientific progress was not exclusive to Europe and Western scientists: for example, the Islamic scientist Ibn al-Haytham published work on experimental methodology to study light and optics centuries before Boyle and Charles employed similar techniques to study gases. It is thought that his contributions laid the groundwork for the development of the scientific method and had a lasting impact on subsequent scientific thinkers in both Islamic regions and later in Europe.

P6.7 Gas pressure and volume

AQA GCSE Physics specification points

4.3.3.2 Pressure in gases

Ⓗ 4.3.3.3 Increasing the pressure of a gas

Smart Curriculum Narrative

Forces and motion

Theme/subtheme

Forces and motion/Pressure

Disciplinary knowledge

- WS 1.2
- MS 3b, 3c

Practical

Gas pressure and temperature

	Learning outcomes		
Learning objectives	**Developing Aiming for 3**	**Secure Aiming for 5**	**Extending Aiming for 8**
Explain in terms of particle behaviour why the pressure of a fixed mass of gas decreases when its volume increases	State that the pressure of a gas increases when it is compressed (at a constant temperature)	**Explain in terms of particle behaviour why the pressure of a fixed mass of gas decreases when its volume increases**	Explain why the temperature and mass of gas must remain constant when investigating pressure volume relationship
Use the relationship $pV =$ constant to calculate the change in pressure after a change in volume	Use the relationship $pV =$ constant to calculate the constant	**Use the relationship $pV =$ constant to calculate the change in pressure after a change in volume**	Solve a variety of problems and calculate either the change in pressure or volume for a fixed mass of gas at constant temperature
Ⓗ Explain why the temperature of a gas increases when it is compressed	State that doing work on a gas increases its internal energy	**Ⓗ Explain why the temperature of a gas increases when it is compressed**	Explain why a gas might cool as it expands

Tier 2 vocabulary	Tier 3 vocabulary
constant, cubic metres, equation, gas, molecule	Boyle's law, fixed mass, force, inversely proportional

Lesson plan

Reactivate knowledge

1 What causes gas pressure?

2 What happens to the pressure of a gas if it is heated?

3 What is internal energy?

Common misconception

Students may think gases have no weight (or mass). Emphasise that mass is a measure of the amount of matter. In each state, the particles of matter don't vanish, they have different amounts of internal energy. Imagine a drop of water is placed in a sealed plastic container (with all air excluded) is warmed until it turns to water vapour. The total mass stays the same, because the number of particles inside the container stays the same.

Answers: 1 Collision of gas particles *2* Increases *3* Sum of the kinetic and potential energies of all the particles of the substance

Trigger interest

Demonstration: Balloon in a syringe (5 mins)

Recap the ideas covered in the previous lesson, which was about the relationship between gas pressure p and temperature T. Point out that another useful property of a gas is the volume V of space it occupies – and that this lesson investigates the relationship between p and V.

Inflate a balloon so it just fits inside a wide syringe. Demonstrate the compression and expansion of the balloon as the pressure changes – carefully keeping the end of the syringe closed and slowly adjusting the barrel. Get students to describe what they observe, focusing on the volume of the balloon and the pressure inside the syringe.

Exposition

Boyle's law (15 mins)

As before, simulations of contained gases are very useful. Focus students' attention on the particle collisions at the container walls, and how these increase as volume decreases. They should form a correct notion of the 'oppositeness' of pressure and volume very quickly.

Introduce Boyle's law as the formal relationship $pV = \text{constant}$. Explain that this means that if p increases, V must decrease at the same rate. Demonstrate this mathematically with a few simple calculations. Explain that this law requires that pairs of values for p and V (for the same gas) should always multiply to give a constant – this will not be explicitly apparent to some students. Ask the students some simple questions so that they can apply this relationship themselves using mini-whiteboards.

H **Doing work on a gas:** Remind students about the concept of doing work and ask them to recall the equation. The idea that work done is equivalent to a transfer of energy is important. Next, ask them what it might mean for work to be done on a gas: demonstrate the effect in real-time using the PhET – Gases Intro simulation, which allows you to reduce the volume (i.e., do work on the gas by applying a force). Ensuring students describe the transfer of energy to the internal energy of the gas, which results in a rise in temperature.

Main activity

Demonstration: Investigating pressure and volume (25 mins)

Frame this demonstration as testing Boyle's law (i.e., pairs of measurements for a gas's pressure p as its volume V changes should always be constant). Emphasising the values of pressure and volume as measurements emphasises the experimental nature of this topic and reinforces the actual technique to verify Boyle's law (i.e., that at least two pairs of readings are required to verify the constant). Students complete the worksheet alongside, or following the demonstration, including graphical considerations of P against V, and P against 1/V.

Extend students by asking how they might go about verifying Boyle's law for a set of data containing many pairs of readings for p and V. They should be comfortable finding a mean value and ignoring obviously anomalous readings.

Further extension can be provided by exploring the role of temperature, and leading students to conclude that Boyle's law only holds if the temperature of the gas stays constant.

Review and practice

Particle summary (10 mins)

Interactive where students fill in the missing words to complete an outline of the key information about particle behaviour during changes in temperature or energy for a material.

P6 Molecules and matter

Introduction to checkpoint intervention

The checkpoint intervention provides suggestions for a lesson to follow up the *P6 Molecules and matter* assessment. Depending on the outcomes of the assessment, these suggestions can be used to move a class towards secure, to target specific learning outcomes, or to consolidate knowledge for students achieving secure outcomes. Use students' outcomes from the checkpoint assessment to address general misconceptions from the content covered. Be prepared to re-cover content as required; three key concepts from the chapter are covered in more detail below.

Checkpoint secure learning outcomes

- Determine the density of a range of solids and liquids.
- Describe the changes in behaviour of the particles in a material during changes of state.
- Use the term 'latent heat' to describe the energy gained by a substance during heating for which there is no change in temperature.
- Calculate the energy required for a change of state using data.
- Describe the relationship between an increase in the temperature of a fixed volume of a gas and the increase in pressure of a gas.
- (H) Explain why the temperature of a gas increases when it is compressed.

Student reflection

Encourage students to reflect on whether there were any checkpoint questions they found difficult or straightforward as well as their preparation for the checkpoint assessment, including how they revised, what they revised, the time spent on revision and what areas they could improve upon. For more strategies, see the Metacognition spread.

Intervention activity: Describe the changes in behaviour of the particles in a material during changes of state.

Simulating particles in matter

Use marbles in a tray to model the states of matter. Any simulation that is tactile will work well, because students will be able to clearly distinguish the rigidity of particles in a solid and the motion of particles in a liquid and gas. This is a useful starting point, to reinforce ideas about internal energy. Next, display large sketches of the three states of matter. Students should be able to identify these comfortably.

Creating a matter flowchart

Arrange the three sketches with space between each sketch to represent the process of melting (solid to liquid) or boiling (liquid to gas). Cut out two short arrows, labelled 'melting' and 'boiling', that can be placed between the sketches. Cut out two long arrows, labelled 'increasing internal energy' and 'decreasing internal energy', that can be placed above or below the three sketches. Students should be able to create a flowchart, describing the motion of particles during each changing of state. Extend this activity by including cut outs to represent heating (e.g., a Bunsen flame) or cooling (e.g., an ice bucket), and ensuring students can verbalise ideas about energy and the forces between particles.

Intervention activity: Calculate the energy required for a change of state using data.

An empty equation triangle

Present students with the equation for energy required to change state, $E = mL$, and ask them to complete an empty equation triangle correctly with the symbols. Then, recap the units of energy and mass, leading them to deduce that specific latent heat, L, must have units of joules per kilogram. This leads nicely into the definition of specific latent heat as the amount of energy required to change the state of one kilogram of a substance. Get students to research some values of L, choosing two examples that show very big and small values. Make sure that they become familiar with what L represents – a value associated with matter changing state.

Applying the equation

Provide simple examples for students to practise applying the equation. Firstly, calculating E using values of m and L, and then rearranging the equation using their equation triangle to calculate m and L. Primarily, this intervention is designed to enable calculations of energy using data, but it is useful to remind students of the principle of conservation of energy. This should frame the idea that something is 'wrong' when energy is supplied to melting ice, but its temperature stays the same. Discuss ideas very simply about weakening (or forming) particle bonds, and the energy required (or expelled) when this occurs during a change of state.

Intervention activity: Describe the relationship between an increase in the temperature of a fixed volume of a gas and the increase in pressure of a gas.

Analysing p, T, and V

Ensure that students have a secure understanding of these quantities, before developing their vocabulary to describe the relationship between them. Use simulations of a gas in a box to explore ideas about each quantity, exaggerating changes to demonstrate what they mean in essence – for example, large and small volumes (the amount of space a gas occupies), high and low temperatures (how hot a gas is, which is related to the average kinetic energy of the particles), and large and small pressures (related to the collisions against container walls).

Describing relationships

With all three symbols displayed, illustrate that the volume is 'fixed' by drawing a box around that symbol. Explain to students that this means we are only considering the relationship between the remaining symbols, p and T, because volume V won't change. With a more secure understanding of what these things mean separately, explore how they depend on each other using a PhET gases intro simulation, increasing the temperature in real-time and observing how pressure changes in response. The first important aspect to draw attention to is the transfer of energy to particles' kinetic stores. The next step – that this results in more frequent collisions with the container walls – should be understood. A proportionality should gradually be acknowledged by the students – which is to say as T increases, so does p, if the volume V stays constant.

P7 Radioactivity

P7.1 Atoms and radiation

AQA GCSE Physics specification points
4.4.1.1 The structure of an atom
4.4.2.1 Radioactive decay and nuclear radiation

Smart Curriculum Narrative
Atomic, nuclear, and quantum physics

Theme/subtheme
Atomic and nuclear physics/Radioactive decay and emissions

Disciplinary knowledge
- WS 4.4
- MS 1a, 1b, 2a

Activity
Investigating radioactivity

	Learning outcomes		
Learning objectives	**Developing Aiming for 3**	**Secure Aiming for 5**	**Extending Aiming for 8**
Describe what is meant by radioactive decay	State that some atomic nuclei are unstable	**Describe what is meant by radioactive decay**	Explain how an unstable nucleus becomes more stable
Describe what is meant by the count-rate	Name a method for detecting radiation	**Describe what is meant by the count-rate**	Explain how radiation is detected
Describe alpha, beta, and gamma radiation	Name the three sub-atomic particles found in an atom (proton, neutron, and electron)	**Describe alpha, beta, and gamma radiation**	Discuss the significance of discovering nuclear radiation

Tier 2 vocabulary	Tier 3 vocabulary
atom, element, energy, particle, randomly, spontaneously, X-ray	alpha particle, background radiation, beta particle, count-rate, detection, emission, gamma, Geiger-Müller tube, instability, ionising, irradiation, nuclear decay, quantum theory, radioactivity, radiation, source, stable, unstable, waves

Lesson plan

Reactivate knowledge

1 Name the particles found inside the nucleus of an atom.

2 Give the charges of the particles found in the nucleus of an atom.

3 What is the charge of an atom?

Common misconception

Students may think that if a human is exposed to a radioactive source they will become radioactive. Explain that radioactivity usually refers to the emission of alpha and beta particles (and sometimes gamma), which is called ionising radiation. Only certain elements are capable of emitting ionising radiation. A human being, or other objects, cannot become radioactive but is contaminated; when humans are exposed to large quantities of ionising radiation, this is called irradiation. The effects are often harmful to living tissue and organs, but the human does not become radioactive.

Answers: 1 Protons and neutrons *2* Protons are positive and neutrons are neutral *3* Atoms are neutral

Trigger interest

Sketching the unknown (5 mins)

Ask students to sketch what they think an atom looks like. Sketches might be a large question mark, since atoms are impossible to image directly.

Use this introduction to build a sense of wonder about the fact that atoms are very mysterious objects, and that the best we can do is to invent models that help us describe how they behave. The most famous model of an atom is to imagine it like a mini Solar System. Discuss what the Sun represents (the nucleus, containing protons and neutrons) and what the orbiting planets represent (electrons). Extend this activity by mentioning that this planetary-model risks various misconceptions about atoms that are not what experiments actually reveal about their structure.

Exposition

A historic accident (10 mins)

This chapter loosely follows the historical timeline of the discovery of the atomic structure. Lean into a storytelling-style approach, illuminating the people behind the developing quantum theory. The mastery of the quantum realm – specifically, the vast energy stored inside the atom – led directly to some of the greatest scientific and technological revolutions of the 20th century, as well as to some of the world's most tragic humanitarian crises. Students should appreciate how leaps in understanding of fundamental physics continue to have an enormous impact on the world stage.

Discuss the accidental discovery of radioactivity by Henri Becquerel, who found out that certain materials could emit highly energetic particles randomly and spontaneously. The young Marie Curie, at the time on the hunt for a thesis topic, would be inspired by this strange new discovery and go on to explore many other radioactive elements.

Main activity

Investigating radioactivity (30 mins)

Show how a Geiger-Müller tube and counter can be used to detect nuclear radiation, starting with a background count. Students can see background detections right from the start of the lesson if set up in advance.

Introduce the cause of this radiation; that unstable nuclei want to get rid of excess energy, which is carried away by radiation. Emphasise the process of moving from nuclear instability to stability as nuclear decay; use diagrams, or an animated video, showing changes to the nucleus itself in order to produce particles detected in the Geiger-Müller tube and counter.

Note the term *radiation* is a generic word that describes things that radiate outwards; in this case, highly energetic particles and waves from the nucleus of an atom. Link this concept to the idea of a count-rate, showing visual representations of each of the three ionising radiation types. Students should research answers to the activity.

Explain that there are both naturally occurring and human-made sources of background radiation, for example, from outer space, bananas, and some rocks.

Review and practice

The champions of quantum theory (10 mins)

Show students an image of the Solvay Conference, 1927, and ask them if they can find Marie Curie. Discuss the prejudice against women and barriers to participation in science at the time. Emphasise her achievements – winner of two Nobel prizes, and instrumental in the development of medical applications of radioactivity – including mobile X-ray units in the First World War and early applications in the treatment of cancer.

Extend students' understanding by asking them to research and list the naturally radioactive elements, along with the dangers these pose to biological tissue (and the effects these had on Marie and her husband, Pierre).

P7.2 The discovery of the nucleus

AQA GCSE Physics specification points

4.4.1.3 The development of the model of the atom (common content with chemistry)

Smart Curriculum Narrative

Atomic, nuclear, and quantum physics

Theme/subtheme

Atomic and nuclear physics/Sub-atomic particles

Disciplinary knowledge

- WS 1.1, 1.2, 1.6
- MS 1b

Activity

The Rutherford model of the atom

	Learning outcomes		
Learning objectives	**Developing** **Aiming for 3**	**Secure** **Aiming for 5**	**Extending** **Aiming for 8**
Describe the plum pudding, the nuclear, and the Bohr model of the atom	Identify the nuclear model of an atom	**Describe the plum pudding, the nuclear, and the Bohr model of the atom**	Compare the plum pudding model, Rutherford model, and Bohr model of the atom in terms of the evidence for each model
Explain how alpha scattering experiments led to the nuclear model of the atom	Describe the evidence provided by the Rutherford scattering experiment	**Explain how alpha scattering experiments led to the nuclear model of the atom**	Explain how the alpha scattering experiment could not be explained using the plum pudding model
Describe how scientists decided there was a positively charged particle called the proton in the nucleus	Name the sub-atomic particles in the model of the atom	**Describe how scientists decided there was a positively charged particle called the proton in the nucleus**	Explain how scientists knew the neutron must exist and why it was not discovered for another 20 years

Tier 2 vocabulary	Tier 3 vocabulary
absorb, atom, particle, reflect, re-emitted, sphere, structure, wave	alpha particle, Bohr model, electron, neutron, nucleus, proton, quantum theory, sub-atomic, ultrasound, X-ray

Lesson plan

Reactivate knowledge

1 What is an alpha particle and what is its charge?

2 Describe the structure of an atom.

3 What happens when two objects with the same charge are brought near to each other?

Common misconception

Students may think that electrons orbit the atomic nucleus like planets orbiting the Sun. Tell students that even though Bohr's 1915 model is a useful tool to develop basic ideas of atoms, physicists' understanding of them have evolved and electrons don't actually move around the nucleus.

Answers: 1 A positively charged particle made of two protons and two neutrons *2* A positively charged nucleus containing protons and neutrons surrounded by negative electrons *3* They repel

Trigger interest

Imaging the unseen (5 mins)

Peel the label off a tin of sponge pudding. Show students the unmarked tin and ask them, in teams, to describe ways they could determine the tin's contents without opening it. Support students with clues: provide images of ultrasounds/X-rays/animals that use ultrasound to build up images, and so on.

The idea that students should develop is that you can build up an image of something by firing it with waves or particles, and studying how they behave subsequently (i.e., after they are absorbed, re-emitted, or reflected).

Exposition

Imagining the atom (15 mins)

Impress upon students the fact that atoms are so unimaginably tiny that they cannot be observed directly. Outline that atoms have been in the realm of human imagination since the times of ancient Greece (Democritus' belief of hard, indivisible spheres) which prevailed until the conception of the plum pudding model; then to the idea of a central nucleus and orbiting electrons by Japanese physicist Hantaro Nagaoka, which he called the Saturnian atom, which was adapted to the now generally accepted Bohr model. Emphasise the global nature of knowledge exchange through history, and that scientific breakthroughs rarely come as a result of isolated endeavour.

Provide a story to outline Geiger and Marsden's experimental work, mathematically analysed by Rutherford, which confirmed Nagaoka's brilliant intuition. Ensure students appreciate the great 'wow-moment' in discovering alpha particles 'bouncing back' off the thin sheet of gold foil. Summarise the key points by displaying a two-column table with the observations and conclusions from this experiment; the separation of observation and conclusion is the key conceptual skill of the lesson.

Next, compare the Rutherford model briefly with the plum pudding model and ask students to propose why this was rejected. Support with animations and simulations as much as possible.

Main activity

The Rutherford model of the atom (25 mins)

Students complete the worksheet to answer comprehension-style questions about the story behind the development of the nuclear and Bohr models. Circulate to ask questions that draw out knowledge and explore ideas that may interest students outside the scope of the specification. For example, ideas about quantum uncertainty, or other non-western narratives of the developing quantum theory.

Review and practice

I don't believe it (10 mins)

Students complete the interactive and then use this summary to write a letter to an unconvinced scientist who wants to hold on to the plum pudding model, and suggest current controversial ideas that persist even in light of overwhelming scientific evidence to the contrary, for example, Moon landings, flat-Earth, etc.

P7.3 Changes in the nucleus

AQA GCSE Physics specification points

4.4.1.2 Mass number, atomic number and isotopes

4.4.2.2 Nuclear equations

Smart Curriculum Narrative

Atomic, nuclear, and quantum physics

Theme/subtheme

Atomic and nuclear physics/Sub-atomic particles

Disciplinary knowledge

- WS 1.4, 4.1
- MS 3c

Activity

Alpha, beta, and gamma emission

	Learning outcomes		
Learning objectives	**Developing Aiming for 3**	**Secure Aiming for 5**	**Extending Aiming for 8**
Determine the number of protons, electrons, and neutrons in an isotope by using nuclear notation	State the definitions of mass number, atomic number, and isotope	**Determine the number of protons, electrons, and neutrons in an isotope by using nuclear notation**	Explain the difference between the isotopes of an element
Describe the changes to a nucleus after either alpha, beta or gamma decay	Describe alpha, beta, and gamma radiation	**Describe the changes to a nucleus after either alpha, beta, or gamma decay**	Compare alpha, beta, and gamma decay
Complete decay equations for alpha and beta decay	Identify the type of decay taking place from a nuclear equation	**Complete decay equations for alpha and beta decay**	Determine the structure of the stable daughter nuclei after a series of decays

Tier 2 vocabulary	Tier 3 vocabulary
atom, decay, emission, isotope	alpha emission, beta emission, decay, gamma emission, neutron, nuclear, nuclear isotope, proton number, random decay

Lesson plan

Reactivate knowledge

1 What is radioactive decay?

2 What is a beta particle?

3 What is a gamma ray?

Common misconception

Students may think that random nuclear decay means it is strange or unlikely. Respond that the way the word 'random' is used in everyday language or slang is very different to how it is used to describe nuclear decay. The nucleus of an atom decays at random, which means that it is impossible to tell exactly when an atom will decay. In fact, physicists can only determine the probability that an atom will decay in a given amount of time.

Answers: 1 When an unstable nucleus emits radiation to become more stable *2* A fast moving electron *3* Electromagnetic radiation from the nucleus

Trigger interest

Passport pictures (5 mins)

Give students a set of images of diverse figures, for example, sportspeople, scientists, politicians, and musicians. They have a set amount of time to carry out some research to complete their passport information, for example, their name, age, height, and country of birth.

Frame this lesson as the scientific analogue for identifying atoms and nuclei, nuclear notation being the equivalent of the atom's passport image.

Extend students' understanding by setting up a task that gets them to propose what the human passport equivalent for a pair of nuclear isotopes might be.

Exposition

Nuclear notation (15 mins)

Show students some examples of nuclear notation, ensuring they can identify the atomic number (proton number) and mass number (nucleon number).

Support can be provided in a variety of ways here: use diagrams of the nuclei next to their symbols, and explore colour coding wherever possible (e.g., red and blue for protons and neutrons respectively). Students could use small coloured balls, marbles, or plasticine, to build their own nuclear models. 3D models often help visualising nuclear structure better than 2D images. Use mini-whiteboards to test changes to atomic and mass numbers in turn, before combinations of both simultaneously.

Once students are comfortable identifying and interpreting nuclear notation, introduce them to simple examples of nuclear decay via alpha, beta, and gamma emission. It helps to keep these processes distinct, but visible throughout the lesson, for example, have them displayed on separate boards or in different areas of a screen. Contrast the changes to the nucleus in both processes, especially that beta decay involves a mysterious change of a neutron to a proton. Videos and simulations will work very well here. Ensure a Periodic Table is provided where appropriate.

Main activity

Alpha, beta, and gamma emission (25 mins)

Students practise interpreting and calculating changes in atomic and mass numbers.

Extension sheets are available where students write and interpret equations to represent nuclear reactions. Discuss with students, where appropriate, the existence of an anti-electron neutrino in beta emission; they could carry out some independent research to discover more about this mysterious particle and the evidence that led to its discovery.

Review and practice

Name that isotope (10 mins)

Interactive where students complete a table describing various isotopes. They need to fill in missing details such as element name, proton number, mass number, and number of electrons.

P7.4 More about alpha, beta, and gamma radiation

AQA GCSE Physics specification points

4.4.2.1 Radioactive decay and nuclear radiation

4.4.2.4 Radioactive contamination

Smart Curriculum Narrative

Atomic, nuclear, and quantum physics

Theme/subtheme

Atomic and nuclear physics/Radioactive decay and emissions

Disciplinary knowledge

- WS 1.4
- MS 1c

Activity

Radiation in action

	Learning outcomes		
Learning objectives	**Developing Aiming for 3**	**Secure Aiming for 5**	**Extending Aiming for 8**
Describe the different penetrating powers of alpha, beta, and gamma radiation through materials and their range in air	Rank the three types of nuclear radiation in order of their penetrating power	**Describe the different penetrating powers of alpha, beta, and gamma radiation through materials and their range in air**	Explain the different penetrating powers of alpha, beta, and gamma radiation
Describe the difference between contamination and irradiation	Describe the process of ionisation	**Describe the difference between contamination and irradiation**	Compare the hazards caused by contamination and irradiation
Apply understanding of the properties of alpha, beta, and gamma radiation to evaluate the best source for a particular use	State some uses of ionising radiation	**Apply understanding of the properties of alpha, beta, and gamma radiation to evaluate the best source for a particular use**	Evaluate the risks of using radioactive materials

Tier 2 vocabulary	Tier 3 vocabulary
charged, element, emission, magnetic field, nuclear, particle, radiation, radioactive, radioactivity, source, symbol	alpha particle, alpha radiation, beta particle, beta radiation, gamma radiation, gamma waves, ionising, irradiation, penetration, radiation

Lesson plan

Reactivate knowledge

1 Which type of radiation is stopped by paper?

2 Which type of radiation has a charge of +2?

3 Which type of radiation has no charge?

Answers: *1* Alpha *2* Alpha *3* Gamma

Common misconception

Students may think that all radiation is harmful. Tell them there is a big difference between ionising and non-ionising radiation. X-rays are ionising and can cause damage to biological tissue. A smartphone, or wi-fi router, or microwave produces non-ionising radiation which does not harm biological tissue. Microwaves do not harm humans unless the power levels are turned up to extremely high levels, like in a microwave oven. In those cases, it's not the radiation itself that is harmful, but the heating effect it produces.

Trigger interest

Too many symbols? (10 mins)

Scientists use a lot of symbols in their work. Students use the interactive to match some symbols they have met so far with what they represent (e.g., elements, equations, etc.). Discuss the reasons that scientists use symbols, including Greek symbols for many formulae and equations in physics.

Extend students by getting them to explore the development of modern physics and the worldwide adoption of English and Greek notation.

Exposition

Penetrating power (15 mins)

Display the symbols for alpha, beta, and gamma radiation and ask students, using mini-whiteboards, to complete the nuclear notation for each type of ionising radiation.

Frame the demonstration as an investigation to determine the nature of mystery radiation emitted by sources X, Y, and Z. Students, in groups, gather clues as the demonstration is delivered – ensuring they observe how the count-rate on the Geiger counter varies with distance from the source, and with intercepting material. With some collaborative research, they should guess the emitted radiation. Discuss and demonstrate the safety measures that must be used when measuring radiation, and the damage caused by ionisation, emphasising that keeping sources at distance is one of the most effective methods. This is a good opportunity to mention irradiation (i.e., being exposed to radiation) vs. contamination (if radioactive material came into direct contact with skin, or if trapped inside the body, e.g., by swallowing a radioactive substance).

Extend students' understanding by discussing the deflection of charged particles by magnetic fields.

If radioactive sources are not available, then a video demonstrating the penetration depths through different materials, and ranges in air, will work well.

Main activity

Radiation in action (20 mins)

Students complete the Working Scientifically sheet to examine the results from an investigation on beta radiation through cardboard and link it to the measurement and control of cardboard manufacture.

When circulating the classroom, ensure to discuss some of the uses of alpha, beta, and gamma radiation, both in industry and instances where (small doses!) of these sources might be found in everyday life, e.g., in bananas, in rocks and soil in certain parts of the UK, etc.

Review and practice

A spilling scenario (10 mins)

Organise students into groups and give them an imaginary source of alpha, beta, or gamma radiation. Ask them to imagine a scenario in which the source was accidentally dropped and lost; they should discuss and propose the steps that would need to be taken by a school to protect the students and minimise damage to the environment.

Support this activity by requiring the use of key words, like irradiation and contamination, and referring to the properties of alpha, beta, and gamma radiation.

P7.5 Activity and half-life

AQA GCSE Physics specification points

4.4.2.1 Radioactive decay and nuclear radiation

4.4.2.3 Half-lives and the random nature of radioactive decay

Smart Curriculum Narrative

Atomic, nuclear, and quantum physics

Theme/subtheme

Atomic and nuclear physics/Radioactive decay and emissions

Disciplinary knowledge

- WS 1.2, 1.4, 1.5
- MS 1c, 3d, 4a

Practical

Activity and count-rate

	Learning outcomes		
Learning objectives	**Developing Aiming for 3**	**Secure Aiming for 5**	**Extending Aiming for 8**
Explain why the activity of a radioactive sample decreases with time	State a definition for the activity of a radioactive source and give its units	**Explain why the activity of a radioactive sample decreases with time**	Compare a physical model of decay with the decay of nuclei, noting the limitations of the model
(H) Find the ratio of a sample remaining after a given number of half-lives	Define half-life in simple terms such as 'the time it takes for half of the material to decay'	**(H) Find the ratio of a sample remaining after a given number of half-lives**	Use the ratio of final to initial activity to determine the age of a sample
Determine the half-life of a sample from given data	Find the half-life of a substance from a graph of count-rate (or nuclei remaining) against time, with support	**Determine the half-life of a sample from given data**	Outline how the age of organic material can be determined by using radioactive dating

Tier 2 vocabulary	Tier 3 vocabulary
decay, emit, exponential, mass, nuclei, percentage, radiation, radioactive, random, rate of change, ratio, stable	decay curve, exponential decay, exponential growth, half-life, radioactive decay, source

Lesson plan

Reactivate knowledge

1 Radioactive decay is described as random – what does that mean?

2 How can you detect radiation?

3 What is count rate?

Common misconception

Students may think that radioactive decay over a half-life means that half of the substance disappears. Clarify that when one half-life of some radioactive substance passes, this means that half of the radioactive nuclei in the source will have decayed and turned into a stable nuclei that does not emit radiation. The source, therefore, undergoes no significant change in mass.

Answers: 1 Cannot predict when it will happen *2* Using Geiger counter/photographic film *3* The number of decays per second detected

Trigger interest

Exponential growth (10 mins)

Introduce students to the surprising properties of exponential mathematics using the following thought experiment: a philosopher places a grain of rice on the first square of a chess board, two on the next, four on the next, and so on. How many go on the final (sixty-fourth) square?

Plot a graph, live on screen, of grains of rice against square number (i.e., 1–64) showing students the resulting exponential growth curve and the surprising final number of grains.

Extend students' understanding by introducing the mathematical idea that exponential growth/decay means that the rate of change is proportional to the size of the growing quantity, that is, the more/less substance there is, the greater/lesser the rate of change.

Exposition

Radioactive decay (10 mins)

Recap the idea of radioactive decay, in terms of a nucleus emitting radiation (in the form of alpha, beta, or gamma) to form a different, more stable nucleus. Use a diagram to show a set of radioactive nuclei (e.g., red circles) decaying to stable nuclei (e.g., green circles). Support students to build quantitative ideas of this decay rate by asking simple ratio and percentage questions about decays in units of time. This should lead to discussions about activity (decay rate) and then count-rate. Expect students to recall that count-rate is measured using a counter connected to a Geiger-Müller tube.

Main activity

Activity and count-rate (30 mins)

In this practical, students work in pairs to investigate a simple model of radioactive decay using dice. The goal is to plot a graph of their results and find a value for the 'half-life' of their decaying die, that is, the roll number at which the number of radioactive die reduces to half the initial amount. This activity works well if students keep their answers secret; once finished, they should be surprised that their answers are all roughly the same, even though rolling dice is a 'random' process. Even more surprising if you give them a prediction at the start of the practical!

Half-life and the random nature of decay (H)

Following the practical activity, students use the extension sheet to try some graphical and ratio calculations.

Extend discussions by asking students to compare the practical model with radioactive decay of nuclei; they should appreciate that the number of dice used is far smaller than the number of atoms in a typical sample. Alternatively, explore ideas about experimental uncertainty in the average value of half-life, and that this can be reduced by increasing the number of dice used (i.e., increasing the likelihood that each pair of students measures the exact same experimental value for half-life). Combining multiple sets of student results data in a spreadsheet will give a curve which is smoother and much closer to the exponential decay curve shape.

Review and practice

Activity and decay (5 mins)

Students use the interactive to put three different decay curves in order of longest half-life.

P7.6 Nuclear radiation in medicine

AQA GCSE Physics specification points

4.4.3.2 Different half-lives of radioactive isotopes

4.4.3.3 Uses of nuclear radiation

Smart Curriculum Narrative

Atomic, nuclear, and quantum physics

Theme/subtheme

Atomic and nuclear physics/Uses and risks of radioactivity

Disciplinary knowledge

- WS 1.5, 1.6
- MS 1b, 2c, 2f

Activity

Nuclear radiation in medicine

	Learning outcomes		
Learning objectives	**Developing Aiming for 3**	**Secure Aiming for 5**	**Extending Aiming for 8**
Describe how nuclear radiation is used to explore organs or destroy unwanted tissue	Name some medical applications for radioactive substances	**Describe how nuclear radiation is used to explore organs or destroy unwanted tissue**	Evaluate the risks of using radioactive materials in medicine
Explain how half-life affects the hazard of a particular radioactive isotope	State that radioactive isotopes have a wide range of half-lives	**Explain how half-life affects the hazard of a particular radioactive isotope**	Discuss the factors that need to be taken into account when selecting a medical tracer for a diagnostic test
Describe sources of background radiation	State that background radiation is the radiation always around you from your environment	**Describe sources of background radiation**	Explain how precautions to reduce exposure to patients and medical staff work

Tier 2 vocabulary	Tier 3 vocabulary
energy, *microwave*, nuclear, radiation, tumour, wave, X-ray	chemotherapy, gamma camera, gamma treatment, ionising, radiotherapy

Lesson plan

Reactivate knowledge

1 Which type of radiation is the least ionising?

2 What is half-life?

3 Two different radioactive isotopes A and B have the same activity. A has a half-life of 20 years and B has a half-life of four hours. How would their activity compare a day later?

Answers: 1 Gamma *2* The time taken for activity to halve or number of radioactive nuclei to halve *3* A would have the same activity and B would have very little activity

Common misconception

Students may think radiotherapy and chemotherapy are the same. Respond that both are types of treatment for cancer.

Trigger interest

Tumours (10 mins)

Find out what students may already know about cancer. For instance, they could each write a fact or two on a post-it note. Refer to prior knowledge they may already have from Biology lessons. Share a sample of responses with the whole class, then ask students to research short answers to the following questions:

(a) What are tumours? (b) How do they arise?
(c) Are they all dangerous?

It is important to clarify here, when discussing the treatments of cancer, that radiotherapy and chemotherapy are not the same. Radiotherapy and chemotherapy are both types of treatment for cancer. Chemotherapy uses medications that are designed to kill or shrink growing cancer cells, whereas radiotherapy kills cancer cells by using beams of high energy waves or particles, for example, X-rays or protons. Both forms of therapy have harmful side effects, depending on the length and strength of therapy required and the individual circumstances.

It may be appropriate to give a disclaimer to the students at the start of this lesson that it is likely one or more people in the room know somebody who is battling cancer or are affected in another way by the effects of cancer. Encourage sensitivity and maturity among the students while promoting the power of modern science and medicine to fight cancer; it may also be appropriate to encourage individuals to ask questions or share concerns privately if they need to.

Exposition

Radioactivity wonders in modern medicine (15 mins)

Show students as many images and videos of tracers and gamma camera images (compared with X-rays) as appropriate. Students can propose ideas of why tracers are useful, and give prompts so they can consider the pros and cons of this procedure. Ensure they conclude correctly why alpha sources cannot be used.

Outline the operation of radiation detectors outside the body, especially in the context of gamma tracers. Damage to healthy cells should be covered, alongside consideration of exposure time beneficially (i.e., implants using alpha or beta radiation, for long-term exposure to tumours) and detrimentally (i.e., over-exposure when not necessary).

Support students with the main activity by providing a summary of key points, including the benefits and risks of each treatment to the patient, along with the precautions taken by medical staff.

Main activity

Nuclear radiation in medicine (20 mins)

Students complete the comprehension-style worksheet to apply ideas about how radiation is used in medicine in a variety of contexts.

It is useful to circulate during this activity, with a tablet or laptop, to show animations and videos relating to the procedures and how treatment is performed as described in the worksheet. These are available with an Internet search; check these in advance of the lesson as some can be quite graphic and/or distressing.

Review and practice

Informed consent (10 mins)

Set up a debate in which students discuss who should have the final say about medical diagnosis and treatment. How much information should be provided to the patient?

Expand this discussion to include children and those unable to give consent, for example, adults with additional needs or mental health conditions.

P7.7 Nuclear fission

AQA GCSE Physics specification points

4.4.3.2 Different half-lives of radioactive isotopes

4.4.4.1 Nuclear fission

Smart Curriculum Narrative

Atomic, nuclear, and quantum physics

Theme/subtheme

Atomic and nuclear physics/Nuclear energy

Disciplinary knowledge

- MS 1b

Activity

A domino chain

Learning objectives	Learning outcomes: Developing Aiming for 3	Learning outcomes: Secure Aiming for 5	Learning outcomes: Extending Aiming for 8
Distinguish between induced fission and spontaneous fission	Describe how nuclear fission is the splitting of a large nucleus to form two smaller nuclei	**Distinguish between induced fission and spontaneous fission**	Explain the differences between naturally occurring isotopes and enriched nuclear fuels
Describe nuclear fission	State that energy is released in a nuclear fission reaction	**Describe nuclear fission**	Explain the differences between naturally occurring isotopes and enriched nuclear fuels
Describe how a chain reaction is controlled in a nuclear power station and what happens if the chain reaction is uncontrolled	Draw a diagram to represent a chain reaction	**Describe how a chain reaction is controlled in a nuclear power station and what happens if the chain reaction is uncontrolled**	Explain how a steady-state induced fission reaction can be maintained

Tier 2 vocabulary	Tier 3 vocabulary
absorbing, induced, nuclear, nuclei, nucleus, spontaneous	chain reaction, control rod, coolant, critical reaction, fission, fusion, nuclear power

Lesson plan

Reactivate knowledge

1 Give the charges of protons and neutrons.

2 Describe the force between two protons.

3 Uranium has a mass number of 235 and atomic number 92. Determine the number of neutrons and protons in the nucleus.

Common misconception

Students may think that fission and fusion are the same thing. Explain that although these words sound very similar, they actually mean the opposite of each other. Fusion occurs when two nuclei join together, whereas fission occurs when a single nucleus splits into two (or more) nuclei.

Answers: 1 Protons are positive and neutrons are neutral *2* Repelling/non-contact force *3* 92 protons and 143 neutrons

Trigger interest

Nuclear power map (5 mins)

Show students a map of nuclear power plants in the UK and ask them to find out how much of the UK's electricity supply came from nuclear power (as of 2020, this was around 16%). Discuss misconceptions around nuclear fission and fusion, emphasising the nearness of international teams to commercialising nuclear fusion.

Recent fusion breakthroughs in this field, and the implications of fusion technology to global energy supply and cost, can be used to inspire discussion and to encourage further independent research.

Exposition

Fission and chain reactions (15 mins)

Allow students to explore an electronic simulation of fission reactions inside a nuclear reactor. They should be able to provide a verbal summary, with prompts, of induced nuclear fission and the chain reaction, identifying the operation of control rods and the coolant.

Ensure students appreciate the difference between spontaneous fission and induced emission and ask what would happen if the process was uncontrolled (i.e., a nuclear weapon), and how this is controlled by absorbing excess neutrons.

Support the activity by providing a list of specific questions and observations. A list of key terms and definitions is a good tool to guide their investigations.

Main activity

A domino chain (25 mins)

Ask students to use this demonstration to distinguish between a critical reaction (one for which the reaction rate stays constant) and a runaway chain reaction (one for which the reaction rate increases).

Set up a simple chain where one domino knocks over another, and another in which one domino knocks over two, two knock over three, and so on.

Students should note that in both cases the reaction must be started off, but then rapidly snowballs. They should then complete the Working Scientifically: *A domino chain* worksheet to explore how chain reactions can lead to devastating effects and how these must be controlled.

Review and practice

Nuclear threat (10 mins)

The story of Oppenheimer and the tragedy at Chernobyl are two useful case studies to explore the moral, social, and political implications of scientific progress. Students should appreciate the difficult role of scientists around the world in pursuing research, appeasing the needs and demands of government (and funders!), whilst upholding social and ethical responsibility.

P7.8 Nuclear fusion

AQA GCSE Physics specification points

4.4.4.2 Nuclear fusion

Smart Curriculum Narrative

Atomic, nuclear, and quantum physics

Theme/subtheme

Atomic and nuclear physics/Nuclear energy

Disciplinary knowledge

- MS 3c

Activity

Fusion reactions

	Learning outcomes		
Learning objectives	**Developing Aiming for 3**	**Secure Aiming for 5**	**Extending Aiming for 8**
Describe nuclear fusion	Give an example of nuclear fusion	**Describe nuclear fusion**	Compare nuclear fission and fusion
Explain why nuclear fusion releases energy	State that nuclear fusion releases energy	**Explain why nuclear fusion releases energy**	Consider the charges on a nuclei and explain why nuclear fusion is difficult
Describe the conditions needed for fusion	State that nuclear fusion is the energy releasing process in the Sun	**Describe the conditions needed for fusion**	Explain why high temperatures and pressures are needed for nuclear fusion

Tier 2 vocabulary	Tier 3 vocabulary
energy, mass, nuclear, nuclei, nucleus, product, reactant, solar storm, temperature	binding energy, carbon-12, fission, fusion, inward gravitational force, main sequence, outward pressure

Lesson plan

Reactivate knowledge

1 What is nuclear fission?

2 Name an element which can undergo fission.

3 How is energy released in nuclear fission?

Common misconception

Students may think that fission and fusion are the same thing. Explain that although these words sound very similar, they actually mean the opposite of each other. Fusion occurs when two nuclei join together, whereas fission occurs when a single nucleus splits into two (or more) nuclei.

Answers: 1 When a large nucleus absorbs a neutron and splits into two smaller nuclei and two/three neutrons *2* Uranium or plutonium *3* Some mass is converted into energy

Trigger interest

The star of the show (10 mins)

Frame this lesson as a story about the Sun, which is home to nuclear fusion. Create a simple true/false quiz to introduce students to some interesting and surprising facts about the Sun, for example, temperature of the surface vs its corona, solar storms, relative size compared to Earth, and so on.

Extend students' understanding by asking open-ended questions to gauge prior knowledge and interest about the Sun and other stars. Broad ideas about the Sun's main sequence, and the balance between inward gravitational forces and outward pressure can be explored in the context of nuclear fusion.

Exposition

Missing mass? (10 mins)

Introduce the puzzle of nuclear fusion, in terms of the mystery of missing mass (being converted to energy). The fact that the mass of the reactants is greater than the mass of the products should be a 'wow moment' that will help students remember the key idea. Explore ideas about where this energy goes – i.e., the kinetic energy of the particles, or radiated away to the thermal energy stores of the surroundings.

Use videos and animations of the fusion process wherever possible. Emphasise the fact that nuclear fusion is the opposite process to nuclear fission, and that both processes release energy.

Discuss the process in the Sun, where there is a readily available source of fusible (i.e., light) nuclei and the right conditions – high temperature and pressure – to cause them to fuse. Contrast these conditions with the design of a fusion reactor on Earth, and how difficult it is to maintain the right conditions within such reactors.

Main activity

Fusion reactions (30 mins)

Students complete the activity describing fusion processes in the Sun.

There are various extension opportunities here: the concept of binding energy, which is linked to the 'missing mass' (or mass defect), can be introduced to explain why there is an energy release.

Groups of students could roleplay the processes of nuclear fusion and fission, while other groups assess their performance based on how well key features of each process are depicted, for example, the conditions for fission/fusion, energy release, and the comparative mass/stability of the reactants and products.

Review and practice

Forged in the stars (5 mins)

"We are all children of the cosmos."

Ask students to discuss and suggest what this statement might mean. Scaffold this question as appropriate, providing hints about the presence of carbon-12 in all life forms on Earth. Students should conclude that the carbon-12 in our bodies was formed in stars through fusion processes and be able to suggest the stages that produce carbon-12 from lighter elements.

P7.9 Nuclear issues

AQA GCSE Physics specification points

4.4.2.4 Radioactive contamination

4.4.3.1 Background radiation

Smart Curriculum Narrative

Atomic, nuclear, and quantum physics

Theme/subtheme

Atomic and nuclear physics/Radioactive decay and emissions

Disciplinary knowledge

- WS 1.5, 1.6, 4.4
- MS 1b, 4a

Activity

Nuclear issues

	Learning outcomes		
Learning objectives	**Developing Aiming for 3**	**Secure Aiming for 5**	**Extending Aiming for 8**
Describe how location and/or occupation can affect your radiation dose	State that radiation dose tells you how much radiation someone has been exposed to	**Describe how location and/or occupation can affect your dose**	Explain how background radiation can be used to assess radioactive risks
Compare the risks and damage associated with alpha, beta, and gamma radiation inside and outside the body	State that there are risks if there is a nuclear accident	**Compare the risks and damage associated with alpha, beta, and gamma radiation inside and outside the body**	Describe and explain the safety precautions that need to take place after a large nuclear accident
Discuss the difficulties associated with the handling and storage of nuclear waste	Describe how nuclear waste is very dangerous and must be stored safely for very long periods of time	**Discuss the difficulties associated with the handling and storage of nuclear waste**	Evaluate in detail a variety of storage or disposal solutions for nuclear waste

Tier 2 vocabulary	Tier 3 vocabulary
element, energy, *microwave*, percentage, pie chart, power, radiation, radioactivity, ultraviolet, X-rays	alpha radiation, background radiation, beta radiation, gamma radiation, ionising, nuclear waste, reactor

Lesson plan

Reactivate knowledge

1 What is background radiation?

2 Name a natural and a human-made source of background radiation.

3 What is half-life?

Common misconception

Students may think that rare elements are the only sources of harmful radiation. Clarify that 'harmful radiation' needs to be qualified. Alpha, beta, and gamma radiation are types of ionising radiation, and can be harmful in certain circumstances, such as proximity and exposure times. However, other forms of radiation, such as ultraviolet, X-rays, and microwaves, are non-ionising but also have harmful effects.

Answers: 1 The radiation in your environment *2* Natural: cosmic rays, rocks, food and drink, air (radon); human-made: nuclear reactors, nuclear weapons testing, X-rays/medical uses *3* The time taken for activity to halve or number of radioactive nuclei to halve

Trigger interest

An empty pie (10 mins)

Show students a pie chart with all the sources of background radiation and their contribution to an average dose removed. Ask students to match background sources to each part of the pie chart. Challenge students to estimate the percentage contributions to the unnumbered chart, and then compare to real figures.

Exposition

All around us (10 mins)

With a complete dataset of background radiation, present students with a map of the UK that shows locations with relatively high levels of background radiation. Discuss typical radiation doses and how this is significantly affected by some medical treatments and occupations. Recap the properties of alpha, beta, and gamma radiation and compare the relative risks of these inside and outside the body. Incorporate a discussion of the handling and disposal of nuclear waste produced by reactors. There is an opportunity here to mention low, intermediate, and high level waste and how each is dealt with.

Main activity

Nuclear issues (25 mins)

Students complete the worksheet exploring the issues surrounding nuclear storage and waste. Encourage independent research wherever possible, and draw out further ideas of the ongoing wider debate about humanity's energy demands and sustainability challenges.

Review and practice

Radioactivity (10 mins)

Students complete the interactive to summarise what they have learned about radioactivity, including the advantages and disadvantages of using radiation and nuclear power in the various contexts introduced.

P7 Radioactivity

Introduction to checkpoint intervention

The checkpoint intervention provides suggestions for a lesson to follow up the *P7 Radioactivity* checkpoint assessment. Depending on the outcomes of the assessment, these suggestions can be used to move a class towards secure, to target specific learning outcomes, or to consolidate knowledge for students achieving secure outcomes. Use students' outcomes from the checkpoint assessment to address general misconceptions from the content covered. Be prepared to re-cover content as required; three key concepts from the chapter are covered in more detail below.

Checkpoint secure learning outcomes

- Describe the plum pudding, the nuclear, and the Bohr model of the atom.
- Complete decay equations for alpha and beta decay.
- Ⓗ Find the ratio of a sample remaining after a given number of half-lives.
- Explain how half-life affects the hazard of a particular radioactive isotope.
- Describe the conditions needed for fusion.
- Compare the risks and damage associated with alpha, beta, and gamma radiation inside and outside the body.

Student reflection

Encourage students to reflect on whether there were any checkpoint questions they found difficult or straightforward as well as their preparation for the checkpoint assessment, including how they revised, what they revised, the time spent on revision, and what areas they could improve upon. For more strategies, see the Metacognition spread.

Intervention activity: Complete decay equations for alpha and beta decay.

Containers are nuclei

Students should be able to complete equations for nuclear decay once the underlying concept is firmly understood, given the required numerical treatment is typically straightforward. Give students, in pairs, small plastic containers holding a set of red and blue marbles, or alternatively, tear coloured paper into small pieces. The containers represent the nucleus of an atom, and the marbles represent protons and neutrons. Create an activity to explore nuclear decay, scaffolding as appropriate, for example, start simply by asking students to remove whole numbers of each nucleon, to get them identifying the correct particle in turn.

Alpha and beta decay

Introduce the concept of an alpha particle as a combination of two red and two blue marbles. Use sticky tape, or sticky tack, to emphasise the fact that the protons and neutrons leave the nucleus together and encourage one of the students to carry it away 'energetically'. This is nuclear decay; students should be ready to translate their roleplayed example into a decay equation, with reminders about the meaning and position of the atomic number and mass number aside the nuclear symbol. Provide a Periodic Table, so they can identify their nuclei, and the resulting nucleus following decay.

Beta decay can be explored in a similar way; one of the students should request an exchange of marbles (neutron for a proton), which triggers their partner to carry a green marble far away, representing the ejected, energetic electron.

The final result should be analysed and transferred to an equation: students should conclude the atomic number is increased, the mass number remains the same, and the new nucleus' identity from the Periodic Table.

Ⓗ Intervention activity: Find the ratio of a sample remaining after a given number of half lives.

Tabulating half-life information

List some examples of radioactive nuclei with a wide range of half-lives, for example, in seconds, days, and years. Ask students, on mini-whiteboards, to draw a set of 32 circles, representing a carbon-14 nucleus (which has a half-life of 5370 years). How many circles remain after one half-life has transpired? With reference to the definition of half-life, expect students to erase 16 circles. Use three student volunteers to tabulate this data, with one student overseeing each of the following information: the number of nuclei-circles (initially 32), the number of half-lives passing n (initially 0), and the fractional amount (or ratio). They should be able to complete the next two or three rows. For example: 16, $n = 1$, and $\frac{1}{2}$ or 50%, and so on.

Understanding the exponent

Verify that students can determine the total number of years, using the number of half-lives passing. For example, ask students how many years have passed if $n = 3$ (this would be $5370 \times 3 = 16\,110$). Then do this in reverse, for example, the number of half-lives given a number of years transpiring (using whole numbers of half-lives, for ease).

Students should be able to use these approaches to determine the ratio remaining after a given number of half-lives; referring back to their table, they will either recognise that $n = 4$ is the row corresponding to a fraction of $\frac{1}{16}$, or even better, that this means $\frac{1}{2} \times \frac{1}{2} \times \frac{1}{2} \times \frac{1}{2} = \frac{1}{2^4}$, which is the best 'shortcut', or multiplying factor, for the initial number of nuclei.

Intervention activity: Describe the conditions needed for fusion.

Overcoming repulsion

Use two student volunteers to act as repelling charged particles. They can either roleplay this directly, acting out the effects of greater repulsion the closer they come together, or alternatively, use large bar magnets to give students a tangible sense of this repulsion (being careful to avoid misconceptions – the magnets, in this case, act as an analogy for positive electric charge!) The fact that like charges repel is the central concept that students should grasp before moving on to describing and justifying conditions for fusion.

The need for pressure and temperature

The hinge question is: *How can we encourage positively charged nuclei to get close enough to fuse?*

Ask students to create a mind map of some ideas – their creativity here will be interesting!

Scaffold their responses by prompting them to think about speed (i.e., leading to ideas about kinetic energy, and therefore, temperature) and particle density (i.e., leading to ideas about being so close to each other so as to increase the chance of a successful collision). Expect students to accept these necessary conditions relatively easily. Finish by asking where in the Solar System we might be able to find an environment with super high temperature and pressure. There are two valid answers to explore: the Sun's core, and inside modern nuclear fusion reactors!

3 Forces in action

Introduction to unit

In this unit, students will build on their prior knowledge and skills in relation to forces and motion. The nature of forces will be developed mathematically, with more emphasis on calculations involving motion, moments, work done, elastic potential energy, and, for higher tier students, momentum, and its conservation. A more sophisticated analysis of graphs and resultant forces will allow students to determine the nature of a body's motion relating to velocity, displacement, and acceleration, and there will be more detail regarding pressure and forces in terms of the random motion of gas particles. Students will become familiar with the workings of gears and levers. Scalar and vector quantities will also be discussed.

Assessment

Suggested minimum guided learning hours: 25 hours

Exam: Paper 2

Prerequisite knowledge from KS3 (chapter references from *Oxford Smart* Activate)

P1.1.1 Introduction to forces, P1.1.2 Squashing and stretching

- Forces are pushes or pulls, arising from the interaction between two objects.
- How to use force arrows in diagrams, adding forces in one dimension
- Forces can deform objects.

P1.1.5 Balanced and unbalanced

- Forces can be balanced and unbalanced.

P2.3.6 Turning forces

- Moment is the turning effect of a force.

P3.2.8 Reach: Elastic energy and Hooke's Law

- Force-extension is a linear relationship, Hooke's Law as a special case.
- Work done and energy changes on deformation.

P3.1.4 Speed and distance-time graphs

- How to calculate speed, distance, and time using the equation *speed = distance ÷ time*.

P3.1.6 Resultant forces

- The effect a force has on an object depends on direction of force and its size.

P2.3.4 Pressure in liquids

- Pressure in liquids increases with depth.

AQA GCSE Physics specification links

4.5 Forces

- 4.5.1 Forces and their interactions
- 4.5.2 Work done and energy transfer
- 4.5.3 Forces and elasticity
- 4.5.4 Moments, levers, and gears (physics only)
- 4.5.5 Pressure and pressure difference in fluids (physics only)
- 4.5.6 Forces and motion
- (H) 4.5.7 Momentum

Identity, awe, and wonder

Students are likely to be interested in sport and it would be worth discussing the nature of forces in sport and whether they need to be large all the time or not. Also, how do the vertical and horizontal forces need to be considered when, for example, causing motion to a ball in rugby, football, golf, or snooker? The horizontal force applied to an object has no effect on its vertical motion and vice versa – how would this knowledge be helpful a sportsman? Many students will be interested in motor vehicles and how fast they can travel. It would be worth asking students why a car will always have a maximum speed and what factors affect this. In terms of forces at a much higher level, students could be asked why it is not possible to travel faster than the speed of light.

Required practicals

AQA required practicals	Practical skills	Topic
Required practical 6: Force and extension Investigate the relationship between force and extension for a spring.	AT 1 – use appropriate apparatus to make and record length accurately. AT 2 – use appropriate apparatus to measure and observe the effect of force on the extension of springs and collect the data required to plot a force-extension graph.	P10.8
Required practical 7: Acceleration Investigate the effect of varying the force on the acceleration of an object of constant mass and the effect of varying the mass of an object on the acceleration produced by a constant force.	AT 1 – use appropriate apparatus to make and record measurements of length, mass, and time accurately. AT 2 – use appropriate apparatus to measure and observe the effect of force. AT 3 – use appropriate apparatus and techniques for measuring motion, including determination of speed and rate of change of speed (acceleration/deceleration).	P10.1

Working scientifically

Students will understand how scientific methods and theories have developed over time, giving us a better understanding of the physical world based on up to date and reliable data, collected from experiments, taking into consideration the importance of peer review. They will also use a variety of models to make predictions and draw conclusions from their results and observations.

Students will plan and carry out investigations, identifying dependent, independent, and control variables. They will consider which apparatus to use, techniques to deploy, and explain why.

Students will collect and analyse data and be expected to use correct scientific vocabulary, quantities, and units in their answers.

Maths skills

Students will be expected to recognise and use fractions, decimals, and percentages in calculations, and represent very large and small numbers in standard form. They will use an appropriate number of significant figures when manipulating numerical data. Students will be able to construct and interpret data graphically, calculate mean values, and state correlations between two variables.

They will be expected to rearrange formulae and substitute values in order to calculate the size of physical quantities, using appropriate units. Students will also calculate areas and volumes, for example when dealing with pressure.

Literacy links

The practical investigations that students will encounter in this unit will serve as an opportunity for writing skills, such as the writing of a step-by-step method. Other literacy-based activities in this unit are included in P8.1, P8.5, P9.4, and P10.4.

P8 Forces in balance

P8.1 Vectors and scalars

AQA GCSE Physics specification points

4.5.1.1 Scalar and vector quantities

Smart Curriculum Narrative

Forces and motion

Theme/subtheme

Forces and motion/Types of force

Disciplinary knowledge

- MS 1a, 2a

Activity

Scalars and vectors

Learning objectives	Learning outcomes: Developing Aiming for 3	Learning outcomes: Secure Aiming for 5	Learning outcomes: Extending Aiming for 8
Explain the difference between distance and displacement	State a definition for scalar and vector quantities	**Explain the difference between distance and displacement**	Explain how an object can have zero displacement but a non-zero distance
Draw a scale diagram to represent a single vector	State that vector quantities can be represented by an arrow	**Draw a scale diagram to represent a single vector**	Interpret a scale diagram to determine the magnitude and direction of a vector
Categorise a wide range of quantities as either a vector or a scalar	List some common scalars and vectors	**Categorise a wide range of quantities as either a vector or a scalar**	Explain why it is easy to combine scalar quantities but harder to combine vector quantities

Tier 2 vocabulary	Tier 3 vocabulary
bearing, direction, distance, horizontal, protractor, *volume*	displacement, magnitude, scalar, *vector*

Lesson plan

Reactivate knowledge

1 Give two examples of a physical quantity (quantity that can be measured).

2 How do you represent a force in a diagram?

3 Why do you represent forces this way?

Common misconception

Students may think that if an object has zero displacement it means it has not travelled anywhere. Tell students that an object can travel a large distance and return to its starting point. Because displacement is a vector, it has a starting point and end point. The displacement will have zero magnitude if the object has the same starting point and end point. This is best illustrated with a diagram, showing the journey of an object that clearly shows its start position and end position.

Answers: 1 Any acceptable quantity; distance, volume, time, force, temperature, current *2* Using an arrow *3* Because they have a direction as well as a size

Trigger interest

Measuring instruments (10 mins)

Spread a variety of measuring instruments around the classroom. Students, in pairs, compete against a timer to find and record the name and quantity measured by each piece of equipment.

Extend this activity by requiring the name, quantity, and unit(s) of measurement, and the uncertainty associated with each measurement.

Exposition

Scalars vs vectors I (15 mins)

Discuss the similarities and differences between distance and displacement. Use plenty of examples, such as those in the Student Book, ensuring that students understand the concepts of magnitude (size) and direction. A local map can be used to highlight and discuss distance and displacement between known locations.

Link back to ideas in the starter activity and discuss some of the instruments in more depth. For example, a measuring cylinder only ever measures a scalar quantity (volume), but a metre ruler and a protractor can be used together to measure displacement.

Using a large interactive grid, or square paper, show how a simple vector (e.g., horizontal and to the right) can be represented by drawing an arrow, labelling its tip, tail, magnitude, and direction.

Scaffold using mini-whiteboards, for example, how long is the vector in terms of the number of squares it covers, or in which direction does it point, building their conceptual understanding gradually with reference to simple scales and units.

For instance, using a scale of 1 square = 1 N, expect students to correctly interpret the magnitude and direction of a 3 N force to the right.

Extend this discussion where appropriate, introducing more complex scales and compass directions/angles and bearings.

Main activity

Scalars and vectors (20 mins)

Students complete the activity sheet to examine, in more detail, the relationship between distance and displacement. The worksheet can then be self or peer marked.

Review and practice

Scalars vs vectors II (10 mins)

Create a poster with the word *scalar* and another with the word *vector*. Place them at opposite ends of the classroom. Students should move between the two posters as you call out different quantities. Students can stay in the middle of the classroom space if they are unsure.

Extend this activity by abstracting quantities wherever possible. For example, calling them out in words, or displaying them as numbers with units, or images/representations of each quantity without additional words or labels.

P8.2 Forces between objects

AQA GCSE Physics specification points

4.5.1.2 Contact and non-contact forces

4.5.6.2.3 Newton's Third Law

Smart Curriculum Narrative

Forces and motion

Theme/subtheme

Forces and motion/Types of force

Disciplinary knowledge

- WS 1.2
- MS 1a, 2a

Activity

Forces between objects

Learning objectives	Learning outcomes: Developing Aiming for 3	Learning outcomes: Secure Aiming for 5	Learning outcomes: Extending Aiming for 8
Describe how forces arise	Describe the action of forces	**Describe how forces arise**	Use scale diagrams to represent the sizes of forces acting on an object
Describe contact and non-contact forces	Give examples of contact and non-contact forces	**Describe contact and non-contact forces**	Compare contact and non-contact forces
Identify and describe Newton's Third Law pairs of forces in different situations	State Newton's Third Law of motion	**Identify and describe Newton's Third Law pairs of forces in different situations**	Explain how objects move when each force acting has an equal and opposite force

Tier 2 vocabulary	Tier 3 vocabulary
force, mass, motion, Newton meter, speed, spring, *stationary*	accelerating, action-reaction pair, constant velocity, contact force, frictional, gravitational, magnitude, Newton's Third Law, non-contact force, *vector*

Lesson plan

Reactivate knowledge

1 Give one effect of a force on an object.

2 Give two examples of a contact force and two examples of a non-contact force.

3 Write down Newton's Third Law.

Common misconception

Students may think that pairs of forces, as described by Newton's Third Law, cancel out because they have equal magnitude and act in opposite directions.

This misconception is usually clarified by identifying if each force acts on the same body and cancel out (which is an example of Newton's First Law of motion), or if each force acts on two separate objects. Using this condition, alongside the fact that each force in a force-pair must be the same type, Newton's Third Law can be verified and applied correctly.

Answers: 1 Any one from: change shape, change speed, change direction *2* Any two from: Contact; friction/air resistance/tension/drag/upthrust. Non-contact; gravitational/electrostatic/magnetic *3* Forces come in equal and opposite pairs/every action has an equal and opposite reaction

Trigger interest

Force diagrams (10 mins)

Provide groups of students with a set of images showing different objects in a variety of states of motion. Students should sort these images under key words (stationary, constant velocity, or accelerating), and mark on the forces they think are acting on the objects. Circulate the groups and observe how students illustrate forces. Draw out good examples of force vectors, especially where shown acting on the point at which the force acts.

Collect a list of forces that students correctly identify, praising the whole class collectively.

Exposition

The size and direction of forces (15 mins)

Demonstrate the action of some forces by pushing and pulling different objects to show that forces can have different magnitudes, and that the direction the force acts in is important. This will establish that forces are vector quantities. Students can drag a few items or hang masses from a spring using a Newton meter to observe different sizes of forces. Charged balloons, and magnets, can be included to explore non-contact forces.

Introduce the idea of the action-reaction pair of Newton's Third Law by using a book at rest on a table as an example. Draw two diagrams – the book only, with a contact force acting upward and from its lower surface; and the table only, with a contact force acting downward from its upper surface. Emphasise that forces always act in pairs, and that each force in this action-reaction pair always act on two different objects. To complete this example, draw a third diagram again showing the book only, with the weight force acting downwards. Use the language of Newton's Third Law (e.g., "the Earth exerts a downwards force on the book, so...") leading students to identify the correct upwards force on the Earth. Draw this fourth diagram of Earth, reinforcing that the pair of forces are the same type (gravitational, non-contact) and equal and opposite.

Support this explanation by highlighting, or colour-coding, the forces, for example, drawing action-reaction pairs in the same style (e.g., dashed and bold lines, or in the same colour). Different types of forces, for example gravitational and frictional, can be visually distinguished in a similar way.

Main activity

Forces between objects (20 mins)

Students read a passage on high diving describing the heights, times, and potential risks involved. They use this passage to identify contact and non-contact forces, action and reaction pairs, and to explain changes in motion due to the action of forces.

Review and practice

Forces between objects (10 mins)

A simple walking-talking demonstration of how Newton's Third Law explains why we are able to walk forwards (or jump upwards!) is a striking way to end the lesson. Students often find it surprising that in order to walk forwards/jump upwards, the entire planet must (in principle) move backwards/move downwards. Discuss this and some other surprising implications of Newton's Third Law, before using the interactive to sort examples of contact and non-contact forces.

Then, students will finish by identifying the action-reaction pairs in the given scenario of a car driving at constant speed.

P8.3 Resultant forces

AQA GCSE Physics specification points

4.5.1.4 Resultant forces

Smart Curriculum Narrative

Forces and motion

Theme/subtheme

Forces and motion/Newton's laws

Disciplinary knowledge

- WS 1.2

Activity

Resultant forces and their effects

	Learning outcomes		
Learning objectives	**Developing Aiming for 3**	**Secure Aiming for 5**	**Extending Aiming for 8**
Describe the effect of zero resultant forces on the motion of moving and stationary objects	State Newton's First Law of motion	**Describe the effect of zero resultant forces on the motion of moving and stationary objects**	Use Newton's First Law to explain why a ball rolling on the floor will slow down
Describe the effect of non-zero resultant forces on the motion of moving and stationary objects	Calculate a resultant force from two parallel forces acting in opposite directions	**Describe the effect of non-zero resultant forces on the motion of moving and stationary objects**	Explain the effect of non-zero resultant forces on the motion of moving and stationary objects
(H) Draw free body diagrams for objects where resultant force is zero and non-zero	Identify all the forces acting on an object	**(H) Draw free body diagrams for objects where resultant force is zero and non-zero**	Draw and interpret free body diagrams

Tier 2 vocabulary	Tier 3 vocabulary
balanced, *constant*, direction, force, mass, maximum, speed, unbalanced	acceleration, free-body force diagram, g-force, Newton's First Law, resultant force, upthrust, *vector*, velocity

Lesson plan

Reactivate knowledge

1 The forces acting on an object are balanced. What will the object do?

2 The forces acting on an object are unbalanced. What will the object do?

3 Force is an example of a vector quantity. What is a vector quantity?

Common misconception

Students may think that continuous motion requires a continuous force. Explain to students that continuous motion requires all forces acting on an object to be balanced (or zero). Imagine pushing a heavy trolley in a supermarket. It needs a continuous pushing-force to keep it moving, otherwise friction would slow it down and cause it to stop moving. In this sense, the pushing-force is required to counterbalance friction. If there were little-to-no friction involved (like on a very smooth surface or icy road), the heavy trolley would slide away and continue moving at a constant speed, because there would be little-to-no overall force acting on it.

Answers: 1 Stay still or move at a constant speed *2* Accelerate/speed up, decelerate/slow down, change direction *3* A quantity that has both direction and size

Trigger interest

Balanced and unbalanced forces (10 mins)

Show the students a toy boat floating on water and ask them to identify and draw a diagram showing the two forces acting on the boat, representing their forces as vectors. Add small masses, one at a time, each time asking students to redraw their vector forces diagram (expecting to see the force vectors increasing but remaining balanced). Keep doing this until the boat sinks. Ask them to draw a final diagram of the forces at the time when the boat was sinking.

Scaffold this activity appropriately – it may be necessary to remind students of the upthrust force, and to reference the fact that this force has a maximum value. Alternatively, use an interactive white-board with pre-drawn vector arrows; students can drag and drop the arrows onto a sketch of the toy boat, and discuss the effect of balanced/unbalanced forces.

Exposition

Newton's First Law (15 mins)

Show students a selection of images and videos that demonstrate Newton's First Law of motion, for example, a cruising aircraft, an asteroid moving at constant velocity through deep space, a stationary object, a floating object, and so on. Ask students to discuss what each of these images have in common, in terms of their motion. Lead students, with prompts and keywords, to the idea that their speed and direction (velocity) is unchanging, which is a result of the fact that the forces acting on them are balanced. Introduce the idea of a resultant force causing acceleration, emphasising that acceleration is a change in velocity (speed and/or direction).

Use a video that shows the extreme acceleration effects to demonstrate the idea, for example, pilots experiencing extreme g-forces. Ask students if they have ever experienced the effects of acceleration and get two volunteers to describe the sensation, sketching the scenario with simple force vectors as they do.

Higher tier candidates should use this opportunity to sketch the scenario described by their peers, on mini-whiteboards, in the form of a free-body force diagram. They need to focus on individual objects, the forces acting on them, and the points from which they act. Emphasise the term *force vector* as opposed to *arrow* when describing the forces.

Main activity

Resultant forces and their effects (20 mins)

Students complete the activity sheet where they apply their knowledge and understanding of zero and non-zero resultant forces on three different objects (a motorbike, an aircraft, and a runner) and carry out calculations of resultant forces from forces applied on the different objects.

Give higher tier candidates the opportunity to interpret additional free-body force diagrams and describe the effects of forces on different objects.

Review and practice

Resultant forces (10 mins)

Students complete the interactive to identify true and false statements about the forces acting on a cyclist.

Higher tier candidates then identify the correct free-body force diagram for a submarine moving forward and descending.

P8.4 Moments at work

AQA GCSE Physics specification points

4.5.4 Moments, levers and gears

Smart Curriculum Narrative

Forces and motion

Theme/subtheme

Forces and motion/Rotation

Disciplinary knowledge

- MS 3c

Practical

Investigating the turning effect of a force

Learning objectives	Learning outcomes: Developing Aiming for 3	Learning outcomes: Secure Aiming for 5	Learning outcomes: Extending Aiming for 8
Describe what the moment of a force measures	Give example of forces causing rotation	**Describe what the moment of a force measures**	Identify everyday examples of moments in action
Perform calculations involving moments, including rearrangement of the equation	Give the factors that affect the size of a moment	**Perform calculations involving moments, including rearrangement of the equation**	Apply the equation for a moment where you have to identify the perpendicular distance to the pivot
Describe levers in action	Identify the key features of a lever; effort, load, and pivot	**Describe levers in action**	Explain how a lever can multiply forces

Tier 2 vocabulary	Tier 3 vocabulary
anti-clockwise, clockwise, direction, force, horizontal, mass, pull, push, rotate, vertical	lever, moment, Newton-metre, pivot, turning effect

Lesson plan

Reactivate knowledge

1 What is a turning force called in physics?
2 Give an example of a turning force in action.
3 How can you increase the turning effect of a force?

Common misconception

Students may think that moments are the same as forces. Clarify to students that moments cause things to turn about a fixed point (or pivot). The size of this turning effect depends on two things – the size of the force applied and how far the force is applied away from the position of the pivot. Hence, moments are measured in Newton-metres, whereas force is measured only in Newtons.

Answers: 1 Moment *2* Any suitable example; any lever, door, tap, steering wheel *3* Increase the size of the force or increase the distance from the pivot

Trigger interest

Right tool for the job (10 mins)

Show students a selection of tools (e.g., crowbar, screwdriver, spanner) and some images that illustrate what they are used for (e.g., opening a box, paint tin, or unscrewing a bolt). For each image, students should vote on which tool they would use for the job and describe how each tool works.

Support how they articulate their responses by providing key words like force, pivot, twist, push, pull, and rotate. There is opportunity to highlight the use of tools in ancient human civilisations, for example, the likelihood that ancient civilisations in England and in Egypt understood the action of forces and levers in order to build, for example, Stonehenge and the pyramids.

Exposition

Levers (10 mins)

Introduce the concept of a moment as the turning effect of a force. To build student confidence around describing these turning effects, and their directions (i.e., clockwise or anti-clockwise), sketch a series of pivoted horizontal levers. Students should determine where to position vertical forces to cause the levers to rotate in either a clockwise or anti-clockwise direction.

Support this introduction by giving pairs of students a metre ruler and some masses to familiarise themselves with pivots, forces, moments, and rotations. Prompt them to form some ideas of the factors that affect the size of the moments, which they will investigate more rigorously in the practical activity. It will be beneficial to model a moment calculation before the practical activity begins, so students can interpret the data they collect.

Main activity

Investigating the turning effect of a force (25 mins)

Students should carry out the investigation to clearly identify the factors that affect the size of the moment (force and distance). The students will design results tables that will allow recording of the force acting on the Newton meter and the weight of the masses and distance from the pivot (provide this, where additional support is required).

Where appropriate, choose pairs or groups in advance and ensure roles are specified and assigned, that is, ensure gender-balance across roles like setting up and packing away, adding and removing masses, and recording results. Highlight and reward examples of good communication and teamwork in pairs and groups.

Following the practical, give students differentiated moment calculations. This can include calculation of the moments produced by the moving masses in their experiment and the force recorded by the Newton meter.

Review and practice

Moments at work (10 mins)

Students complete the interactive to calculate the moment applied to a nut using a spanner. They also identify the gears from a bicycle that have the greatest moment.

As a challenge, ask students to keep a record of the variety of ways in which they apply a moment (e.g., opening a door). They should aim to come to the next lesson with at least five ways.

P8.5 More about levers and gears

AQA GCSE Physics specification points

4.5.4 Moments, levers, and gears

Smart Curriculum Narrative

Forces and motion

Theme/subtheme

Forces and motion/Rotation

Disciplinary knowledge

- MS 3c

Practical

Wheels and axles

	Learning outcomes		
Learning objectives	**Developing Aiming for 3**	**Secure Aiming for 5**	**Extending Aiming for 8**
Describe the action of levers being used as force multipliers	Identify levers being used as force multipliers	**Describe the action of levers being used as force multipliers**	Explain why a force multiplier requires the effort force to move through a larger distance than the load
Describe the action of a pair of gears in terms of increasing or decreasing the moment of a force	State that gears can be used to increase or decrease the moment of a force	**Describe the action of a pair of gears in terms of increasing or decreasing the moment of a force**	Apply moment equations to a gear system to determine force on driven gear or moment of driven gear
Investigate the action of a set of two gears	Observe the relative speed and directions of a pair of gears of different diameters	**Investigate the action of a set of two gears**	Analyse the results for systems of gears of different ratios and draw conclusions

Tier 2 vocabulary	Tier 3 vocabulary
force, gear, lever, mass, rotation, speed	axle, moment, multiplier, Newton-metre, pivot, turning effect

Lesson plan

Reactivate knowledge

1 What is a lever?

2 What do levers do?

3 Levers can be used as force multipliers. What is a force multiplier?

Common misconception

Students may think that moments are the same as forces. Clarify to students that moments cause things to turn about a fixed point (or pivot). The size of this turning effect depends on two things – the size of the force applied and how far the force is applied away from the position of the pivot. Hence, moments are measured in Newton-metres, whereas force is measured only in Newtons.

Answers: 1 A rigid bar that rotates about a pivot *2* They transmit rotational effects of forces *3* A lever where the load is larger than the effort

Trigger interest

Elephant vs mouse (10 mins)

Put students in groups to suggest an answer to the following puzzle:

A elephant has a mass of 2000 kg, and a mouse has a mass of 10 g. If they sit on either side of a seesaw, and the elephant is 0.5 m from the pivot, how far away should the mouse sit to balance the seesaw?

This can lead to a calculation of the moment from the elephant and then the distance the mouse needs to be from the pivot (100 km).

Extend this by asking students to suggest the simplifications in this question: *Could a seesaw with a length of 200 km be constructed?*

Exposition

Levers vs gears (10 mins)

For the following activities, demonstrate using equipment or online simulation wherever possible.

Use the devices shown in the Student Book to explain the action of levers being used as force multipliers.

Following this, demonstrate the action of a pair of gears on the moment of a force. Emphasise that the changes in distance from the shaft (pivot) allow changes in the sizes of the moment, linked to changes in the speed of rotation.

Main activity

Wheels and axles (25 mins)

Students will explore force multipliers using a wheel and axle combination.

Where appropriate, choose pairs or groups in advance and ensure roles are specified and assigned, that is, ensure gender-balance across roles like making and building axle components, adding and removing masses, and recording results. Highlight and reward examples of good communication and teamwork in pairs and groups.

Review and practice

Archimedes' lever (10 mins)

Provide students with this quote from the Greek philosopher, Archimedes:

'Give me a lever long enough and a fulcrum [pivot] on which to place it, and I shall move the world.'

In groups, students discuss and interpret this statement, using ideas about levers and force multipliers from the lesson.

P8.6 Moments and equilibrium

AQA GCSE Physics specification points

4.5.4 Moments, levers, and gears

Smart Curriculum Narrative

Forces and motion

Theme/subtheme

Forces and motion/Rotation

Disciplinary knowledge

- MS 3a, 3c

Practicals

Measuring an unknown weight

Measuring the weight of a beam

	Learning outcomes		
Learning objectives	**Developing Aiming for 3**	**Secure Aiming for 5**	**Extending Aiming for 8**
Use calculation of moments to determine if a seesaw is in equilibrium	Calculate moments using the appropriate equation	**Use calculation of moments to determine if a seesaw is in equilibrium**	Use calculations to determine if an object with three or more moments is in equilibrium
Apply the principle of moments to a balanced object to determine an unknown force or distance	State that the total clockwise moments about a pivot equals the total anti-clockwise moments about a pivot if the object is in equilibrium	**Apply the principle of moments to a balanced object to determine an unknown force or distance**	Apply the principle of moments in a range of contexts to find an unknown force or distance, including ones where there are three or more moments
Design a method and use calculations to find the weight of an object	Follow instructions to find the weight of an object using a balanced beam	**Design a method and use calculations to find the weight of an object**	Establish the possible range of uncertainty of a weight using repeat values

Tier 2 vocabulary	Tier 3 vocabulary
anti-clockwise, balanced, clockwise, direction, equilibrium, force, inequality, mass, mean, Newton	moment, Newton, Newton-metre, pivot, turning effect, vector

Lesson plan

Reactivate knowledge

1. Define a moment.
2. State the law of moments.
3. Describe how an adult can balance a seesaw with a child.

Common misconception

Students may think that equilibrium means that no forces are acting. Explain that equilibrium means that the sum of all forces is zero, or, in other words, that all the forces acting on an object cancel out. Since forces are vectors, both size and direction matter – so equally sized forces can cancel out if they act in opposite directions on the same object. Similarly, moments can cancel out to leave an object in equilibrium, depending on their relative magnitude and direction.

Answers: 1 Moment = force × perpendicular distance from the pivot to the line of action of the force *2* When an object is in equilibrium/balanced, the sum of the clockwise moments = sum of the anti-clockwise moments *3* Adult has to sit much closer to the pivot than the child

Trigger interest

Moments and equilibrium (5 mins)

Interactive where students answer questions on moments and a balanced seesaw. Circulate and ask students to explain their reasoning – some will be using moment calculations already whilst others will be using their intuition. Validate and praise both lines of reasoning. Allude to the usefulness of a general framework for balancing forces and moments by considering and comparing these quantities on both sides of a pivot.

Exposition

A balancing act (15 mins)

Display the principle of moments on a large board, as well as the equation for moments, which will follow directly from the interactive. Use a diagram of a simple pivoted beam, showing arrows to denote clockwise motion and anti-clockwise motion.

Lead the students through some differentiated example calculations to ensure they can apply the mathematical techniques, starting with relatively simple situations (e.g., just one force on each side of the pivot).

It is useful to provide a consistent calculation framework, for example,

clockwise moments = anti-clockwise moments

emphasising the left and right side of the equal sign. Sticking each force and distance in brackets often makes the maths with three or more moments clearer for the students if they are challenged to do this.

Anticipate students challenging the use or relevance of this physics (i.e., it's not just about seesaws!) Provide them with real-life applications of the principle of moments (e.g., large cranes used for construction and engineering) or challenge them to do so, for example, by listing one or two job roles that may require an understanding of moments and equilibrium. Alternatively, provide students with a short case study, e.g., a structrual engineer. Students could suggest the types of projects they might work on (suspension bridges, sports stadiums, skyscrapers) and the types of skills required for their role. Link this back to the need for an understanding about forces, equilibrium, and the importance of problem-solving.

Main activity

Practicals (30 mins)

Two practical activities are provided here. Support these experiments by providing a ready-made table for repeat readings, to prompt students appropriately. Challenge individuals or groups to express their final weights as a range, using inequality signs.

Measuring an unknown weight

Students set up the apparatus as shown in the Student Book. They balance the masses and measure distances from the pivot, in order to find the unknown mass. They should repeat the process and find the mean mass, checking their answers using a top-pan balance.

Measuring the weight of a beam

Students balance a beam by placing it on the pivot, then rebalance it by adjusting the position of added mass. This should be repeated several times with the pivot in different positions. Emphasise the importance of careful positioning to gain a precise answer in this experiment.

Review and practice

The odd one out (5 mins)

Show students a diagram of four seesaws, three of which are balanced and one of which is unbalanced (and about to topple). The task is for students to find the odd one out; they will need to perform calculations. Ask students to vote as pairs using mini-whiteboards. Circulate and provide clues as needed, referring to their use of the principle of moments (comparing clockwise and anti-clockwise moments) as introduced earlier in the lesson.

Ⓗ P8.7 The parallelogram of forces

AQA GCSE Physics specification points

4.5.1.4 Resultant forces

Smart Curriculum Narrative

Forces and motion

Theme/subtheme

Forces and motion/Types of force

Disciplinary knowledge

- WS 1.2
- MS 5a, 5b

Practical

Parallelogram of forces

	Learning outcomes		
Learning objectives	**Developing Aiming for 3**	**Secure Aiming for 5**	**Extending Aiming for 8**
Use vector diagrams drawn to scale to find the resultant of two forces acting at right angles to each other	Draw a vector diagram to scale for two forces acting at right angles to each other	**Use vector diagrams drawn to scale to find the resultant of two forces acting at right angles to each other**	Use vector diagrams to determine the resultant of two forces
Use the parallelogram of forces to find the resultant of two forces acting at an angle	Describe how you find the resultant force using the parallelogram of forces	**Use the parallelogram of forces to find the resultant of two forces acting at an angle**	Analyse a wide range system of non-parallel forces using a parallelogram technique
Draw a vector diagram for recorded measurements	Measure forces and angles for three weights in equilibrium	**Draw a vector diagram for recorded measurements**	Investigate non-parallel forces acting on a system in equilibrium to verify the parallelogram of forces

Tier 2 vocabulary	Tier 3 vocabulary
diagonal, direction, force, mass, parallelogram, perpendicular, Pythagoras' theorem	resultant, *vector*

Lesson plan

Reactivate knowledge

1 Are forces scalar or vector quantities? Explain your answer.

2 What is the resultant force?

3 Describe how a 10 N force and 6 N force can have a resultant force of 4 N.

Common misconception

Students may think that the resultant force between two forces is a third, additional force. Tell students that the resultant force that acts on an object is the overall effect of two (or more) forces. There is no third, or additional force acting on the object. This is best illustrated by showing two equal and opposite forces acting on an object; it should be agreeable that there are two, and only two forces acting – but the overall, or resultant force on the object, is zero.

Answers: 1 They are vector as they have both magnitude and direction *2* A single force that has the same effect as all the forces acting on an object *3* The 10 N and 6 N force act in opposite directions (10–6 = 4 N)

Trigger interest

Resultant force quiz (5 mins)

Provide students with a few simple resultant force questions (forces acting on the same line), which they answer on their mini-whiteboard. Gradually increase the difficulty, for example, by introducing three or four forces, and then a final question where two forces are perpendicular.

Students should appreciate the key difference between this final question and the others – that is, the forces acting along different lines. The hinge question then becomes: *How do we add forces together to work out the resultant force when they act along different lines?*

Students could be challenged to use Pythagoras' theorem to solve the final question (e.g., two perpendicular forces of 3 N and 4 N).

Exposition

Parallelograms (10 mins)

Recap students' knowledge of parallelograms using mini-whiteboards, ensuring they can recognise pairs of parallel lines. The key idea is the diagonal of the parallelogram; scaffold how students recognise and label this diagonal and emphasise its importance.

Draw a large sketch of a ship and two tugboats, as in Figure 2 of the Student Book. Student volunteers should label the direction of the forces provided by each tugboat. Emphasise the practical use of resultant forces, as the overall effect on the huge ship (being pulled forwards) is greater than the sum of its parts (the individual forces provided diagonally by the two tugboats). Students should recall, therefore, that forces are vectors and both size and direction are important.

Finally, if the two forces from the tugboats are two sides of a parallelogram, ask a volunteer to complete the missing edges of this parallelogram. Label the main diagonal clearly, explaining that this is the resultant force and emphasising that this parallelogram enables us to determine how large it is and its direction.

Main activity

Parallelogram of forces (35 mins)

Students investigate the parallelogram of forces using a pulley and mass arrangement, drawing a scale diagram to determine the resultant force vector.

To support students where needed, each mass can be assumed to represent 1 N and be drawn to a scale of 1 cm to simplify the diagrams.

Review and practice

Finding the resultant force (5 mins)

Interactive where students put in order the instructions for finding the resultant of two forces.

H P8.8 Resolution of forces

AQA GCSE Physics specification points

4.5.1.4 Resultant forces

Smart Curriculum Narrative

Forces and motion

Theme/subtheme

Forces and motion/Representing forces

Disciplinary knowledge

- WS 1.2
- MS 5b

Practical

Testing an incline

	Learning outcomes		
Learning objectives	**Developing Aiming for 3**	**Secure Aiming for 5**	**Extending Aiming for 8**
Draw a scale diagram to resolve a single force into two perpendicular components	State that when you resolve a force you draw two perpendicular components whose resultant is equal to that force	**Draw a scale diagram to resolve a single force into two perpendicular components**	Resolve a pair of forces into the overall perpendicular components
Determine if an object is in equilibrium by considering the horizontal and vertical forces	State the conditions for an object to be in equilibrium	**Determine if an object is in equilibrium by considering the horizontal and vertical forces**	Determine if an object is in equilibrium by considering the horizontal and vertical components of forces
Investigate the effect of increasing the weight of an object on a slope on the component of the weight acting along the slope	Describe how to measure the component of weight parallel to a slope	**Investigate the effect of increasing the weight of an object on a slope on the component of the weight acting along the slope**	Plan a detailed investigation into the effect of increasing the gradient of a slope on the component of the weight acting along the slope

Tier 2 vocabulary	Tier 3 vocabulary
force, horizontal, parallel, perpendicular, slope, vertical	component, frictional, Newton-metre, resolve, *vector*

Lesson plan

Reactivate knowledge

1 State Newton's First Law.

2 Describe how you find the resultant of two forces acting at right angles to each other.

3 Why should you use a sharp pencil when drawing a scale vector diagram?

Common misconception

Students may think that resolving a force means there are two additional forces. Clarify that just like components of a machine are the parts that make it up, the components of a force make up the force itself. These components are not new or additional forces; components help us interpret how much of the force acts in one direction, and how much of the same force acts in a perpendicular direction.

Answers: 1 When forces act on an object, it will remain at rest or keep moving at constant speed in a straight line *2* Draw the forces to scale, then complete the parallelogram, the resultant is the diagonal of the parallelogram *3* Thick lines will make the measurements inaccurate

Trigger interest

Force matchmaking (10 mins)

Prepare a tray of prose descriptions (e.g., *a 20 N force acting at 30 degrees to the horizontal*) on small pieces of paper, and matching sketches. Have students select either prose or sketch and give them 30 seconds to find their corresponding partner. Both students must agree and stick their prose and sketch to a board, writing their names so you can discuss and analyse examples with the class.

Emphasise that the point behind the activity is to be able to describe forces precisely. Additional challenge is provided by using descriptions and sketches of forces on slopes.

Exposition

Resolving vectors (10 mins)

Demonstrate the idea of breaking up a force vector into horizontal and vertical components. It may be useful to refer to a simple coordinate system; expect students to be familiar with coordinates as 'made up of' a certain amount along the x-axis and another along the y-axis (e.g., *along the corridor and up the stairs*). This provides a scaffold for students to grasp the idea that force vectors are similarly made up of two perpendicular components.

Be aware that the shift to resolving forces along a slope is a subtle but considerable conceptual leap for most students. Most simple examples will introduce resolution along the horizontal and vertical (e.g., x and y axes), whilst slopes require resolution with respect to the slope itself. It is useful to point this out explicitly to students, contrasting the two examples side-by-side. Transparent grids to put over diagrams can help with slopes, as the grid can be aligned to the slope. Differentiate additional resolution examples as appropriate.

Keep a large diagram of both versions (resolution along horizontal and vertical axes vs along a slope) displayed to refer back to throughout the lesson. It is worthwhile to draw components as dotted lines to remind students that these comprise the force itself (they are not additional, separate forces).

Main activity

Testing an incline (25 mins)

Students use this experiment to see that the force along a slope (i.e., the component of weight parallel to the slope) increases with the weight of the trolley.

Be aware of misconceptions that may arise during this experiment – emphasise that the force provided by the Newton meter counteracts the parallel component of the weight force. A diagram to refer to will be useful. There are also frictional effects that will need to be considered (i.e., it is likely that doubling the weight does not exactly double the component forces).

Review and practice

Examining equilibrium (10 mins)

Present students with the conditions for a system to be in equilibrium. Give groups a selection of diagrams and ask them to determine if the system is in equilibrium, which they must present and justify to the class.

Differentiate scenarios as appropriate – some groups might have an object with only two equal and anti-parallel forces, while others will need to resolve forces and compare vertical and horizontal components.

P8 Forces in balance

Introduction to checkpoint intervention

The checkpoint intervention provides suggestions for a lesson to follow up the *P8 Forces in balance* checkpoint assessment. Depending on the outcomes of the assessment, these suggestions can be used to move a class towards secure, to target specific learning outcomes, or to consolidate knowledge for students achieving secure outcomes. Use students' outcomes from the checkpoint assessment to address general misconceptions from the content covered. Be prepared to re-cover content as required; three key concepts from the chapter are covered in more detail below.

Checkpoint secure learning outcomes

- Draw a scale diagram to represent a single vector.
- (H) Draw a scale diagram to resolve a single force into two perpendicular components.
- Describe the action of levers being used as force multipliers.
- (H) Use the parallelogram of forces to find the resultant of two forces acting at an angle.
- Determine whether an object is in equilibrium by considering the horizontal and vertical forces.

Student reflection

Encourage students to reflect on whether there were any checkpoint questions they found difficult or straightforward as well as their preparation for the checkpoint assessment, including how they revised, what they revised, the time spent on revision, and what areas they could improve upon. For more strategies, see the Metacognition spread.

Intervention activity: Draw a scale diagram to represent a single vector.

Scales and ratios

Build students' confidence with scales and ratios using interactive or paper maps to trigger interest. The straight line distance between two landmarks could be used as a puzzle for students to solve, given a simple scale as a ratio, for example, 1 cm : 1 km.

Gauge whether students are more comfortable using these scales and ratios in words, for example:

If 1 cm on the map represents 1 km in reality, what distance would 2 cm represent in reality?

Drawing scale diagrams

Once students can manipulate and interpret simple scales and ratios, introduce the key concept that the size of a force can be represented by a scaled vector, just like the size of distances can be represented by a scaled map. Give students a simple scale of 1 cm : 1 N to practise scale drawings on cm-squared paper (horizontal or vertical force vectors), and gradually increase the difficulty of the scale, for example, 1 cm = 0.5 N.

Ⓗ Intervention activity: Draw a scale diagram to resolve a single force into two perpendicular components.

Recognising vectors and components

Students need to recognise a vector having both size and direction, and that it contains two components. This can be explored using PhET – Vector Addition. Students can drag and rotate vectors in a 2D plane, varying magnitude and direction, before visualising the components (helpfully indicated as dotted lines). Scaffold questions as appropriate, for example, create two vectors of equal magnitude, in opposite directions, identify the vertical component, and so on.

Scale and resolution

Only introduce the process of resolution and scaling once students are confident in identifying and comparing vectors and components. Give students a variety of sketched vectors, in different orientations, with unresolved, partially resolved, and fully resolved examples. Students can arrange/sort/complete these as appropriate. The final stage is to interpret and draw vectors to scale, for which a simple scale can be employed (e.g., 1 cm = 1 N). Students will be more comfortable interpreting the magnitude of single forces drawn as horizontal lines, e.g., 3 cm = 3 N, and so on, before resolving and measuring components directly. It may also be useful to, finally, give students practice at drawing vectors to specified angles – e.g., 40° to the horizontal. Model using a protractor to build confidence as appropriate.

Ⓗ Intervention activity: Use the parallelogram of forces to find the resultant of two forces acting at an angle.

Procedure vs interpretation

The procedure of using a parallelogram should be kept distinct from interpreting what the resultant force means. Ensure that students are familiar with parallelograms – first by identifying sets of parallel lines, and then being able to identify parallelograms from a selection of shapes (including a variety of quadrilaterals). Once students can confidently do this, they should be able to identify the longest diagonal within the parallelogram.

The resultant force

Create two vectors out of paper or card and join them together so that they can rotate like hands on a clock, for example, using a split pin. This enables you and students to generate a wide variety of parallelograms. Show students how two forces acting at an angle to each other creates the parallelogram by drawing a parallel line to each vector in turn. The resultant force should be clearly labelled, for example, using a coloured pen.

P9 Motion

P9.1 Speed–time and distance–time graphs

AQA GCSE Physics specification points

4.5.6.1.1 Distance and displacement

4.5.6.1.2 Speed

4.5.6.1.4 The distance–time relationship

Smart Curriculum Narrative

Forces and motion

Theme/subtheme

Forces and motion/Motion

Disciplinary knowledge

- MS 1a, 1b, 1c, 1d, 3b, 3c, 4a, 4b, 4d

Activity

Distance–time graphs

	Learning outcomes		
Learning objectives	**Developing Aiming for 3**	**Secure Aiming for 5**	**Extending Aiming for 8**
Use and rearrange the speed equation to determine either speed, distance, or time	Estimate typical speeds for walking, running, and cycling	**Use and rearrange the speed equation to determine either speed, distance, or time**	Perform calculations of speed, distance, and time which involve conversion to and from SI base units
Use the gradients of distance–time graphs to calculate the speeds of objects	Describe how the gradient of a distance–time graph represents the speed	**Use the gradients of distance–time graphs to calculate the speeds of objects**	Compare the motion of objects by extracting data from a distance–time graph
Describe the motion of an object by interpreting distance–time graphs	Draw distance–time graphs from measurements	**Describe the motion of an object by interpreting distance–time graphs**	Sketch distance–time graphs to represent descriptions of motion or given data

Tier 2 vocabulary	Tier 3 vocabulary
average, distance, fastest, gradient, slowest, speed, time	distance–time graph

Lesson plan

Reactivate knowledge

1 Name some units of speed.

2 Give the equation for calculating speed.

3 How can you tell the speed of an object from a distance–time graph?

Common misconception

Students may think that if an object has a large acceleration, it has a large speed. Students may struggle to distinguish between acceleration and speed (or velocity). An object can have a very slow speed (e.g., a snail) or a very large speed (e.g., an asteroid flying through deep space). However, they could have the same acceleration if, for example, their speed increases by 1 m/s every second. Emphasise the units of acceleration as 'the *change* in m/s... every second', to reinforce the fact that acceleration is the rate of change of speed (or velocity).

Answers: 1 m/s, km/h, mph 2 speed = distance/time 3 The slope of the graph tells you the speed; the steeper the slope the faster the speed

Trigger interest

Lightning Bolt (10 mins)

Show a video of Usain Bolt's world record achievement for the 100 m or 200 m. Ask students what two pieces of information would be needed to calculate Usain's incredible speed. Expect that students have prior knowledge of the speed equation and ask a volunteer to use the information from the race to calculate Usain's speed. Extend, if appropriate, to allude to the fact that this would be his average speed (because his speed changes over the course of the race).

Next, show a collage of moving objects and ask students to put them in order from fastest to slowest. This activity can be differentiated in a variety of ways: students could estimate the speeds before ordering the objects, or speeds could be provided but in different units (e.g., cm/s, m/s, km/h) to prompt students to appreciate units of speed, or they could be provided with data to work out the speeds directly. Examples could be: a worm (0.5 cm/s), a human walking (0.5 m/s), a human running (2 m/s), bicycle (5 m/s), car (20 m/s), passenger aircraft (200 m/s), or the international space station (7 700 m/s). Students can compare a 'typical' running speed to Bolt's record breaking speed.

Exposition

The speed equation (15 mins)

Present the speed equation and emphasise units of metres for distance and seconds for time. Support students using a simple calculation frame, or equation triangle, especially for rearrangement of the equation where needed.

Present examples and get students to try the equation as quickly as possible. Use situations that expose students to typical speeds (walking at 1.5 m/s, running at 3 m/s, and cycling at 6 m/s) as students are expected to recall these values.

Present a table of measurements, like the one shown in Table 1 of the student Book, for the distance at various times of a moving object (e.g., comparing two 100 m runners in intervals of 10 m, showing their respective times side-by-side). Plot this data on a screen, including a line of best fit, so students can quickly see how the data creates a distance–time graph. They can discuss and vote on which line represents the faster sprinter; draw out reference to the gradient of the line.

Main activity

Distance–time graphs (20 mins)

Students complete the activity sheet where they analyse a distance–time graph of a motorbike, describing when it was travelling at the slowest speed, and extracting simple data from the graph to find speed in different phases of motion.

Review and practice

A driving story (10 mins)

Put students into groups and provide each group with a mini-whiteboard. Then present a short paragraph describing your imaginary car journey to school. Forewarn students that the story will contain key pieces of information about speed. Make the story humorous – the aim is to encourage engagement while developing students' collaboration and ability to identify useful pieces of information.

The challenge is to sketch a distance–time graph of the described motion, and to interpret it to demonstrate how it matches the description given.

Support can be provided by highlighting the sentences that correspond to important phases of motion, or alternatively, by giving students the graph and asking them to translate it to prose.

P9.2 Velocity and acceleration

AQA GCSE Physics specification points

4.5.6.1.3 Velocity

4.5.6.1.4 The distance–time relationship

Smart Curriculum Narrative

Forces and motion

Theme/subtheme

Forces and motion/Motion

Disciplinary knowledge

- MS 3b, 3c, 3d, 4a, 4b, 4c, 4d, 4f

Activity

Acceleration

	Learning outcomes		
Learning objectives	Developing Aiming for 3	Secure Aiming for 5	Extending Aiming for 8
H Explain with examples why an object moving in a circle has constant speed but changing velocity	Describe the difference between speed and velocity using an appropriate example	**H Explain with examples why an object moving in a circle has constant speed but changing velocity**	Explain using Newton's First Law of motion why the resultant force is non-zero for an object moving in a circle
Calculate the acceleration of an object using the change in velocity and time	Give the equation relating change in velocity, acceleration, and time	**Calculate the acceleration of an object using the change in velocity and time**	Rearrange the acceleration equation to determine initial velocity, final velocity, or time
Estimate values for everyday accelerations, such as a car slowing down to stop, with help	State that an object slowing down is decelerating	**Estimate values for everyday accelerations, such as a car slowing down to stop, with help**	Estimate values for everyday accelerations

Tier 2 vocabulary	Tier 3 vocabulary
average, distance, negative, positive, speed	acceleration, circular motion, conical pendulum, deceleration, directional, displacement, *vector*, velocity

Lesson plan

Reactivate knowledge

1 What is meant by a vector quantity?

2 What is the difference between distance and displacement?

3 What is acceleration?

Common misconceptions

Students may think that if an object has a large acceleration, it has a large speed. Students may struggle to distinguish between acceleration and speed (or velocity). An object can have a very slow speed (e.g., a snail) or a very large speed (e.g., an asteroid flying through deep space). However, they could have the same acceleration if, for example, their speed increases by 1 m/s every second. Emphasise the units of acceleration as 'the *change* in m/s... every second', to reinforce the fact that acceleration is the rate of change of speed (or velocity).

Answers: 1 Vector quantities have both size and direction *2* Distance is a scalar quantity and displacement is a vector quantity *3* Changing speed

Trigger interest

Getting nowhere fast (10 mins)

Naomi Schiff once set a lap record when she drove a full circuit of a 3 km racetrack in 90 seconds. In her new role as F1 commentator, she describes that her total displacement was 0 m at the end of the lap, and her average speed was 33.3 m/s.

Ask students to discuss answers to the following questions: *How far did Naomi travel around the racetrack? What does she mean by her 'total displacement' being 0 m? How did she calculate her average speed?*

Use this scenario to explain the difference between distance and displacement.

Exposition

Defining velocity and acceleration (10 mins)

Explain the definition of velocity in words, before introducing the equation and formal ideas of vector and directional properties. For example, *just as speed is defined as the change in distance per second, velocity is defined as the change in displacement per second.* Students can struggle with the difference between speed and velocity. Use plenty of examples involving the description of the direction of motion (e.g., forwards and backwards using positive and negative signs). Number lines can be useful here for both displacement and velocity.

Higher tier candidates will need to recognise and interpret circular motion in the context of changing velocity and be able to link this to the presence of a resultant force in terms of Newton's Second Law. A demonstration using PhET – Ladybug Revolution provides a useful visualisation of the changing velocity vector for an object moving in a circle.

Main activity

Acceleration (30 mins)

Students may restrict acceleration to only mean 'getting faster'. Use this as a starting point to explore additional statements, for example, 'slowing down', 'changing speed', 'changing velocity', and 'changing direction' as equivalent conditions satisfying acceleration. Introduce deceleration as another way of saying 'slowing down', which means it is a specific type of acceleration. Introduce the equation and give several example calculations, including examples of starting or end velocity of zero (e.g., at rest, stopped, stationary). Draw attention to the units of acceleration, interpreted as 'metres per second, every second'.

Students complete the worksheet to gain confidence using the equation and interpreting changing velocity and values of acceleration.

Extend this task with the Working Scientifically: *Acceleration* worksheet, in which students explore a method using light gates and a data logger to measure acceleration.

Review and practice

Estimating acceleration (5 mins)

Create a matching activity between some images of everyday examples of acceleration, including deceleration, and values. Examples could include a falling apple, a launching rocket, a rollercoaster, and a car stopping at traffic lights. Alternatively, students can answer relevant exam-style questions from the Student Book.

P9.3 More about velocity–time graphs

AQA GCSE Physics specification points

4.5.6.1.5 Acceleration

Smart Curriculum Narrative

Forces and motion

Theme/subtheme

Forces and motion/Motion

Disciplinary knowledge

- MS 3b, 3c, 4a, 4b, 4c, 4d, 4f

Practical

Investigating acceleration

	Learning outcomes		
Learning objectives	**Developing Aiming for 3**	**Secure Aiming for 5**	**Extending Aiming for 8**
Use the gradient of a velocity–time graph to determine the acceleration	Describe the motion shown on a velocity–time graph	**Use the gradient of a velocity–time graph to determine the acceleration**	Sketch velocity–time graphs to represent descriptions of motion or given data
Ⓗ Calculate the distance travelled using information taken from a velocity–time graph for one section of motion	Identify the feature of a velocity–time graph that represents the distance travelled (the area beneath the line), and compare these values	**Ⓗ Calculate the distance travelled using information taken from a velocity–time graph for one section of motion**	Ⓗ Calculate the total distance travelled from a multi-phase velocity–time graph
Use a series of repeat measurements to find an accurate measurement of the acceleration of a moving object	Measure the acceleration of an object as it moves down a ramp	**Use a series of repeat measurements to find an accurate measurement of the acceleration of a moving object**	Evaluate an experiment into the acceleration of an object in terms of precision based on the spread of repeat measurements

Tier 2 vocabulary	Tier 3 vocabulary
constant, decreasing, gradient, increasing, negative, positive, rest, speed	acceleration, decelerating, velocity, velocity–time graph

Lesson plan

Reactivate knowledge

1 How do you calculate the gradient of a graph?

2 What does a horizontal line on a velocity–time graph represent?

3 How do you calculate acceleration?

Answers: 1 Draw a triangle and the gradient is the height ÷ base *2* Constant speed *3* Acceleration = change in velocity/time taken for the change

Common misconception

Students may think that acceleration is always in the same direction as an object's motion. Discuss the directional property of acceleration in relation to velocity. This is borne out in examples of decelerating objects, for example, a train slowing to a stop in a station. Drawing the direction of motion (the velocity vector) and the acceleration vector (in the opposite direction) will illustrate that this is a case of an object decelerating, or slowing down.

Trigger interest

Speeding up or slowing down? (10 mins)

On mini-whiteboards, ask students to sketch the velocity–time graph of a car that increases its velocity from rest to 10 m/s, and then for the same car that decreases its velocity from 10 m/s to rest.

Scaffold questions around these two foundational examples, such as positive and negative gradient comparisons, and straight or curved lines.

Extend this discussion by asking students to determine the acceleration in both cases if the change in velocity and time is annotated on each axis. Some students could sketch a graph from scratch, given an accurate description of the acceleration.

Exposition

Braking (10 mins)

Recap the definition of acceleration and its associated equation. Present a large velocity–time graph of a car braking, with two distinct sections: travelling at constant velocity, then decelerating to a stop. Labelling or colouring these sections, divide the students into pairs and ask them to choose one of the sections to discuss and interpret.

Circulate and provide prompts about the areas of the sections highlighted – for example, multiplying velocity and time should produce a displacement (or distance). Higher-tier candidates can be given some illustrative values to practise simple calculations of area – including summing the rectangular and triangular area to determine the total distance travelled.

Main activity

Investigating acceleration (25 mins)

Students carry out the investigation depending on the equipment available.

The motion sensor apparatus provides the simplest method; the light-gate method requires additional calculations of speed, from which acceleration can be deduced. This method suits groups that are very comfortable distinguishing between speed and acceleration.

Review and practice

Matching motion (10 mins)

Students use the interactive to match velocity–time graphs with the descriptions of a car's motion. They then interpret a velocity–time graph of an athlete during a training session.

Extend this activity by asking students to read data from the graphs (such as initial and final speed) and to attempt to calculate the acceleration.

P9.4 Analysing motion graphs

AQA GCSE Physics specification points

4.5.6.1.4 The distance–time relationship

4.5.6.1.5 Acceleration

Smart Curriculum Narrative

Forces and motion

Theme/subtheme

Forces and motion/Motion

Disciplinary knowledge

- MS 3b, 3c, 4a, 4b, 4c, 4d, 4f

Activity

Velocity–time graphs

	Learning outcomes		
Learning objectives	**Developing Aiming for 3**	**Secure Aiming for 5**	**Extending Aiming for 8**
Calculate speed, acceleration and distance travelled from data taken from motion graphs	Draw and interpret the appropriate motion graphs from data	**Calculate speed, acceleration, and distance travelled from data taken from motion graphs**	Calculate the distance travelled by counting the squares under the velocity–time graph
(H) Draw a tangent to determine the speed at that point on a distance–time graph of an accelerating object	Identify acceleration on a distance–time graph	**(H) Draw a tangent to determine the speed at that point on a distance–time graph of an accelerating object**	(H) Use a tangent to determine the speed of an object from a distance–time graph
Use the equation $v^2 - u^2 = 2as$ in calculations where the initial or final velocity is zero	State that there is an equation ($v^2 - u^2 = 2as$) to determine constant acceleration if time is not known	**Use the equation $v^2 - u^2 = 2as$ in calculations where the initial or final velocity is zero**	Apply and rearrange the equation $v^2 - u^2 = 2as$ in calculations involving change in velocity and acceleration where both velocities are non-zero

Tier 2 vocabulary	Tier 3 vocabulary
gradient, motion, *stationary, tangent*	acceleration, distance–time graph, velocity, velocity–time graph

Lesson plan

Reactivate knowledge

1 What does a horizontal line on a distance–time graph represent?

2 What does the gradient of a distance–time graph represent?

3 What does the gradient of a velocity–time graph represent?

Answers: 1 Stationary *2* Speed *3* Acceleration

Common misconception

Students may think that acceleration is always in the same direction as an object's motion. Discuss the directional property of acceleration in relation to velocity. This is borne out in examples of decelerating objects, for example, a train slowing to a stop in a station. Drawing the direction of motion (the velocity vector) and the acceleration vector (in the opposite direction) will illustrate that this is a case of an object decelerating, or slowing down.

Trigger interest

Graph matching (5 mins)

The students have to match the descriptions of moving objects with distance–time and velocity–time graphs. Provide three different descriptions of journeys and three graphs that represent the movement for the students to match them with.

This can prove uniquely challenging for some students; the cognitive processing required to identify, interpret, and explain mathematical abstracts (graphs) and wordy descriptions, and then to make links between them, is considerable. Chunk the task appropriately to provide support or differentiate examples wherever possible. For instance, some pairs or groups can be tasked with sorting distance–time and velocity–time graphs, so they're solely focusing on the quantities represented by each axis. Graphs may have only one, or multiple stages of motion, providing layers of challenge as appropriate.

Exposition

Ⓗ Using distance–time graphs (15 mins)

Lead the students through the process of determining the gradient of a line using the tangent. Ensure students can clearly identify when an object is speeding up and when it is slowing down and when an object is stationary (a tangent with a gradient of zero). Students should find some velocities from example graphs to ensure they have mastered the technique.

Referring to a simple sketch of a linear graph with positive gradient, draw out students' prior knowledge of slopes and gradients. Discuss finding the acceleration, comparing a gradient calculation with the equation for acceleration. Students should recognise that change in velocity is represented by the *y*-axis, and change in time is represented by the *x*-axis. Make sure that students are aware of the difference between a velocity–time graph and a distance–time graph.

Introduce the equation of motion, $v^2 - u^2 = 2as$, as an alternative tool to calculate acceleration when time is not known. Give a few simple examples, emphasising that this equation is provided on the physics equation sheet, and that students only need to know how to use it. This is quite a challenging equation to use, and so will need careful modelling where appropriate.

Main activity

Velocity–time graphs (25 mins)

Students complete the worksheet to practise applications of the equations and general interpretations of motion graphs.

Extend higher tier candidates using the tangent techniques to find the acceleration when objects are not accelerating uniformly. Alternatively, students could be challenged to derive the equation $v^2 - u^2 = 2as$ from a velocity–time graph. Shade the areas as depicted in Figure 3 of the Student Book.

Review and practice

Using graphs (10 mins)

Discuss the features of a distance-time or velocty-time graph for Usain Bolt's 100 m race (there are many simple examples available online) linking back to P9.1. Higher tier candidates should note that a tangent to the d-t curve would determine his instantaneous speed.

Students then complete the interactive to answer a series of true or false statements on velocity, then identify velocity–time graphs.

Higher tier candidates can be challenged to correct the false statements.

P9 Motion

Introduction to checkpoint intervention

The checkpoint intervention provides suggestions for a lesson to follow up the *P9 Motion* checkpoint assessment. Depending on the outcomes of the assessment, these suggestions can be used to move a class towards secure, to target specific learning outcomes, or to consolidate knowledge for students achieving secure outcomes. Use students' outcomes from the checkpoint assessment to address general misconceptions from the content covered. Be prepared to re-cover content as required; three key concepts from the chapter are covered in more detail below.

Checkpoint secure learning outcomes

- Use and rearrange the speed equation to determine either speed, distance, or time.
- Use the gradients of distance–time graphs to calculate the speeds of objects.
- Calculate the acceleration of an object using the change in velocity and time.
- (H) Draw a tangent to determine the speed at that point on a distance–time graph of an accelerating object.
- Calculate speed, acceleration, and distance travelled from data taken from motion graphs.
- Use the equation $v^2 - u^2 = 2as$ in calculations where the initial or final velocity is zero.

Student reflection

Encourage students to reflect on whether there were any checkpoint questions they found difficult or straightforward as well as their preparation for the checkpoint assessment, including how they revised, what they revised, the time spent on revision, and what areas they could improve upon. For more strategies, see the Metacognition spread.

Intervention activity: Use the gradients of distance–time graphs to calculate the speeds of objects.

Understanding gradient

Calculating the gradient of a graph is a mathematical skill that students will need to carry out. They will also need to be able to interpret what the gradient means. Sketching a person walking along a hilly path, ask students to imagine how that person would describe the steepness of different slopes as they walk along them. This will encourage them to develop a sense of gradient in more relatable contexts.

Gradient of distance–time graphs

Students should appreciate that calculating a number to describe the steepness of a slope is useful. Discuss the idea that the steepness of a distance–time graph represents the speed of an object, because speed is given by the change in distance divided by the change in time. Provide students with a set of data to plot the graph of a moving object. This should be the simple scenario, for example, an object starting from rest and moving with constant speed, with the distance travelled given in intervals of five or ten seconds.

Model how they can calculate the gradient of the line, ensuring you use the entire line and emphasising the change in the y-value divided by the change in the x-value. As a final step, compare lines of different gradients (the value of the gradient could be given, with or without units of speed) and ask students to identify which lines represent objects moving with faster or slower speeds.

Intervention activity: Calculate the acceleration of an object using the change in velocity and time.

Defining acceleration

Create a card-sorting activity for which students must arrange short sentences about acceleration into columns for 'True', 'False', or 'Not sure'. Sentences should aim to recap core ideas about acceleration, for example, acceleration means change in speed over time, units of acceleration are m/s^2, a body accelerates if its speed is constant but its direction changes, and so on. Supplement this activity with short video clips of objects accelerating – speeding up and slowing down. All clips should have a clear theme of changing speed (velocity) over a period of time.

Calculating acceleration

If students are confident at recognising and defining acceleration, provide a simple calculation framework (e.g., an equation triangle) to help them practise simple calculations. For the first set of questions, it may be helpful to model writing out the known quantities (e.g., velocity and time), and substituting these values explicitly into the equation for acceleration. As confidence grows, introduce different units for velocity and time, and finally, questions requiring rearrangement (e.g., to determine change in velocity).

Intervention activity: Calculate speed, acceleration, and distance travelled from data taken from motion graphs.

Recognising properties of graphs

Create a variety of sketched graphs, for example, straight horizontal and sloped lines, positive and negative gradients, shaded areas, and so on. Spend some time asking students, using mini-whiteboards, to identify and interpret key aspects of motion graphs, for example, quantities and units on each axis, whether the graph shows positive or negative gradient. This will give an early sense of students' basic grasp of recognising and interpreting key quantities from graphs.

Interpreting properties from motion graphs

Shift attention from sketches to graphs plotted on grids, using data, and recap the key quantities and how they can be determined (speed from the gradient of a distance–time graph, and acceleration from the gradient of a velocity–time graph). Plenty of modelled examples using simple scenarios (e.g., uniform speed and acceleration) will give students procedural practice. For higher tier candidates, reason that the area under a velocity–time graph must represent distance (or displacement) by reference to the equation *velocity × time = distance*. A graph of an object moving at constant velocity, so that the area underneath the line is a rectangle, will help support this explanation. Once this is securely understood and interpreted, move onto changing velocity – for example, a triangular graph, such that students can discuss the object's average velocity. The distance travelled is still the area under the line, which in this case is given by the formula for the area of a traingle, $\frac{1}{2}$ base x height.

P10 Forces and motion

P10.1 Force and acceleration

AQA GCSE Physics specification points
4.5.6.2.2 Newton's Second Law

Smart Curriculum Narrative
Forces and motion

Theme/subtheme
Forces and motion/Newton's Laws

Disciplinary knowledge
- MS 1d, 3a, 3b, 3c

Required Practical 7
Acceleration

Learning objectives	Learning outcomes: Developing Aiming for 3	Learning outcomes: Secure Aiming for 5	Learning outcomes: Extending Aiming for 8
Use and rearrange the $F = ma$ equation to find the resultant force acting on an object, its mass, or its acceleration	Describe the effect of changing the mass or the resultant force acting on an object on the acceleration of that object	**Use and rearrange the $F = ma$ equation to find the resultant force acting on an object, its mass, or its acceleration**	Calculate the acceleration of an object acted on by several forces
(H) Explain motion in terms of inertia	Describe how objects require a non-zero resultant force to change their velocity	**(H) Explain motion in terms of inertia**	(H) Define the inertial mass of an object in terms of force and acceleration
Combine separate experimental conclusions of how resultant force affects the acceleration of a mass to form an overall conclusion	Write a conclusion for an investigation into how resultant force affects the acceleration of a mass	**Combine separate experimental conclusions of how resultant force affects the acceleration of a mass to form an overall conclusion**	Evaluate an experiment of how resultant force affects the acceleration of a mass by identifying sources of error and determining uncertainty in the resulting force

Tier 2 vocabulary	Tier 3 vocabulary
acceleration, *constant*, *force*, *mass*, velocity	gravity, inertia, resultant force

Lesson plan

Reactivate knowledge

1. What is acceleration?
2. What is the resultant force?
3. What happens when the resultant force acting on an object is non-zero?

Common misconception

Students may think that to keep an object moving in a straight line, a force has to be continuously applied or it will stop. Tell students that a force will only change an object's motion, for example, by changing its speed or direction. Therefore, an object will continue in a straight line unless acted on by a force.

Answers: 1 Rate of change of velocity 2 A single force that has the same effect as all the forces acting on an object 3 Object accelerates/changes speed/changes direction

Trigger interest

Does the moon also fall? (10 mins)

Give a brief biography of Isaac Newton, his contribution to modern science, especially his use of mathematics to explain how and why things move. Newton asked, "If the apple falls, does the moon also fall?" Realising the answer was "Yes!", he was able to formulate his theories of motion and gravity.

Highlight research that suggests Newton was influenced by the insights of others before him, including many European and Islamic trailblazers. For example, Ibn Al-Haytham, known for his work on optics, may have been one of the 'giants' Newton was referring to when he said, "If I have seen further it is by standing on the shoulders of giants."

Exposition

Resultant forces and acceleration (15 mins)

Use an animation or simulation to demonstrate the meaning of force and acceleration. Ensure that stationary objects, and objects moving at constant velocity, are used to contrast their motion to acceleration. Students could use mini-whiteboards to vote whether an object has resultant forces acting on it, or not, given a demonstration of its motion.

Introduce the equation $F = ma$ as Newton's Second Law. Provide a few simple examples showing how to determine the resultant force F, and ones that require rearrangement to determine m and a. Use an equation triangle as support where necessary.

Emphasise intuition to check whether answers seem sensible. For example, a larger resultant force F is required to accelerate a trolley full of groceries (large m), than if the trolley was empty (small m).

Higher tier candidates should extend this discussion by focusing on relationships between, and definitions of, each symbol.

Using the mathematical relationships $a \propto F$ and $a \propto \frac{1}{m}$, ask students to combine these two statements to reach the expression $a \propto \frac{F}{m}$. This leads to the more familiar $F = ma$

Students should appreciate that F and a are both vectors and they act in the same direction. Introduce the concept of inertia – short clips showing interesting inertial effects work well, for example, pulling a tablecloth – and the definition of m as inertial mass, a measure of how difficult it is for the force F to change its velocity.

Main activity

Acceleration (25 mins)

Students carry out two investigations for the required practical. First, investigate the effect that changing the force on an object has on its acceleration. Then, investigate the effect that changing the mass of the object has on the acceleration when a constant force is applied.

For support, it may be useful to display the three variables as symbols, F, m, and a, and ask students to identify the symbol being held constant. This will enable them to verbalise which variables are being tested and which is the control.

If time permits, students can use both investigations to reach two conclusions to combine into a mathematical relationship. Alternatively, split the class into two groups, one testing each relationship, and then share findings.

Review and practice

Something's wrong (5 mins)

Show a video of a train decelerating. Display: "Objects always move in the direction of the resultant force." Students discuss why this is incorrect, possibly drawing a free-body diagram. Show the corrected phrase: the train moves right, but the resultant force left causes deceleration.

P10.2 Weight and terminal velocity

AQA GCSE Physics specification points

4.5.1.3 Gravity

Smart Curriculum Narrative

Forces and motion

Theme/subtheme

Forces and motion/Motion

Disciplinary knowledge

- MS 2c, 2g, 3a, 3b, 3c, 4c

Practical

Investigating falling

	Learning outcomes		
Learning objectives	**Developing Aiming for 3**	**Secure Aiming for 5**	**Extending Aiming for 8**
Calculate the weight of objects using their mass and the gravitational field strength	State the difference between the mass of an object and its weight	**Calculate the weight of objects using their mass and the gravitational field strength**	Apply the mathematical relationship between mass, weight, and gravitational field strength in a range of situations
Apply the concept of balanced forces to explain why an object falling through a fluid will reach a terminal velocity	Identify terminal velocity on a velocity-time graph	**Apply the concept of balanced forces to explain why an object falling through a fluid will reach a terminal velocity**	Explain the motion of an object for each part of a velocity–time graph of an object falling through a fluid by considering the forces acting through all stages
Investigate the relationship between the mass of an object and the terminal velocity	Investigate the motion of an object when it falls	**Investigate the relationship between the mass of an object and the terminal velocity**	Design preliminary experiments to determine a suitable range for the independent variable and suitable values for the control variables

Tier 2 vocabulary	Tier 3 vocabulary
accelerate, *constant*, fluid, *force*, *mass*, *matter*, speed, velocity, weight	air resistances, descent, freefall, gravitational field, terminal velocity

Lesson plan

Common misconception

Students may think that mass and weight are alternative words for the same property. Mass can be thought of as the amount of 'stuff' (i.e., matter) an object contains. Weight, however, is the force on an object due to the gravitational field it happens to be in. Extend this discussion to the weight on an object on Earth and the Moon.

Reactivate knowledge

1 Give one difference between weight and mass.

2 Draw a force diagram for an object falling at the moment just after you let go of it.

3 What happens to a moving object if the resultant force is zero?

Answers: 1 Appropriate suggestions include: weight measured in Newtons and mass in kg *2* One force arrow – weight acting downwards *3* Moves with constant velocity

Trigger interest

Freefall analysis (5 mins)

Show students a clip of a sky dive, ideally with speed displayed. Ask students to describe the motion of the skydiver as the speed increases and note the point at which the speed remains constant. Identify this as the point in the fall at which acceleration stops, or is zero.

Then ask students to describe how the acceleration changes. Since change in speed is very large at the beginning of the fall, acceleration is also large (maximum). Students should be able to explain that speed changes in smaller increments as the descent continues, so acceleration is decreasing to zero.

Exposition

Weight, air resistance, and terminal velocity (15 mins)

Display an image of a skydiver in descent. Refer to this throughout the lesson for a visual reference of the two acting forces of weight and air resistance.

Ask students to describe what is meant by 'air resistance'. Explain this in terms of the skydiver's body pushing particles out of the way as they fall through the atmosphere. Therefore, there is an upward force on them due to the air (or atmosphere). Draw and label this upward force on the displayed image.

Refer to the misconception, introducing the equation and definition of gravitational field strength, and some calculations of weight for different masses. Students should determine the weight of the skydiver, and then draw and label this downward force on the displayed image.

Finally, tell students that the weight remains constant through the skydiver's descent (since m and g are constant). Ask students to discuss this concept and vote whether the upward air resistance stays constant.

Scaffold their discussion while circulating the room and encourage differences of opinion, providing clues where appropriate. Students should realise that air resistance increases as the skydiver's speed increases and recognise that the forces become balanced when they reach terminal velocity. It may be useful to display three or four copies of the falling skydiver, showing the weight on each image, and the slowly increasing air resistance as they fall.

Main activity

Investigating falling (25 mins)

Students investigate the effect of different masses on the time for a parachute to descend a set height. If time is limited, individual groups could investigate one mass, and then collate results as a class. As there will be considerable variability in the results, the students should focus on repeat measurements to find mean values, reducing the effects of random error. Ensure practical roles are appropriately distributed amongst groups.

Parachutes need to be dropped from a sufficient height to make them effective. A height of >3 m is ideal.

Support students by linking their practical back to the skydiver as frequently as possible; circulate groups and ensure they can describe the motion of their parachute in terms of weight and air resistance being unbalanced or balanced.

Review and practice

Ball bearings are like skydivers (10 mins)

In a tall measuring cylinder, compare a selection of objects (e.g., a coin, a die, different-sized ball bearings) falling through a viscous fluid such as washing up liquid. Pen markings at equal intervals of distance, or equally-spaced rubber bands around the cylinder, may help to analyse the motion more clearly. Slow motion video analysis can be helpful.

Students predict which object will reach terminal velocity first, justifying their predications in terms of the downward force of weight and the upward force of fluid resistance.

P10.3 Forces and braking

AQA GCSE Physics specification points

4.5.6.3.3 Factors affecting braking distance 1

4.5.6.3.4 Factors affecting braking distance 2

Smart Curriculum Narrative

Forces and motion

Theme/subtheme

Forces and motion/Motion

Disciplinary knowledge

- WS 1.5, 2.2.
- MS 1a, 1c, 1d, 2c, 2f, 2h, 3b, 3c

Practical

Reaction time challenge

Learning objectives	Learning outcomes		
	Developing Aiming for 3	Secure Aiming for 5	Extending Aiming for 8
Describe the factors that affect thinking distance and braking distance	Define the terms stopping distance, thinking distance, and braking distance and describe how they are related	**Describe the factors that affect thinking distance and braking distance**	Explain the relative effects of changes of speed on thinking and stopping distance
Explain the dangers of large decelerations	Describe how work done by the friction force of the brakes transfers energy from the kinetic energy store of the car	**Explain the dangers of large decelerations**	Use $F = ma$ to estimate the braking forces for vehicles on the road
Determine the thinking distance using reaction time and initial velocity	Describe one method to measure human reaction time and state typical range of values	**Determine the thinking distance using reaction time and initial velocity**	Use data to calculate acceleration or braking distance

Tier 2 vocabulary	Tier 3 vocabulary
acceleration, *constant*, decelerate, *force*, *mass*, percentage, speed, velocity	air resistance, braking distance, friction, stopping distance, thinking distance

Lesson plan

Reactivate knowledge

1 What is kinetic energy?

2 How do you calculate work done?

3 What is reaction time?

Common misconception

Students may confuse thinking distance and reaction time. Establish that even though both are distinguished by their names, 'distance' and 'time', this misconception is common. Thinking distance is the distance covered by a vehicle while the driver is thinking (the brain responding to a prompt to apply the brakes), whereas reaction time is the time taken to react to the prompt – e.g., the time between seeing something in the road, and applying the brakes.

Answers: 1 Energy an object has because of its movement *2* Work done = force × distance *3* The time it takes to react to a stimulus

Trigger interest

Stop! (10 mins)

Introduce the idea of stopping distance by showing a short clip of a moving vehicle braking to a stop. Provide students with a selection of images to initiate a discussion about how stopping distances can be impacted. For example, fatigue, alcohol, drugs, phones, icy or wet road conditions, tyre conditions, fast moving cars. Explain that car accidents are a common source of serious injury and death, especially among young people, and encourage them to take car safety seriously.

Exposition

Forces on a car (15 mins)

Show a car and discuss the forces acting on it whilst it is moving at a constant velocity. A slow-motion clip will work well, allowing live annotations and commentary. Reinforce the idea that forces must balance if the car is not accelerating, identifying the resistive forces (friction and air resistance) clearly.

Students should understand the factors affecting the overall stopping distance, but they need to be clear which factors affect the thinking and braking distance.

For higher tier candidates, the braking distance should be linked closely to the decelerating forces using $F = ma$, for example, a maximum braking force results in a maximum deceleration, and so a faster car will require more time (and distance) to stop. Example calculations should be given and finding stopping distances are provided on the Bump up your grade worksheet.

Main activity

Reaction time challenge (20 mins)

Demonstrate this practical a few times with a student volunteer, including use of the equation $t = \sqrt{\frac{2d}{g}}$ to determine reaction time.

Students should be familiar with the method before beginning, and work in pairs. Reaction time in this practical simulates the time it takes to react and begin braking in a moving vehicle (see Exposition).

One student holds a metre ruler, so that the base of the ruler is just above the other's hand. They drop the ruler suddenly and the other student has to catch it, between thumb and index finger. The distance the ruler falls before it is caught can be used to work out the reaction time. Ensure that the equation to do this is provided or displayed.

Students should repeat the process multiple times and determine an average reaction time. Illustrate the effects of distraction by having a conversation with students as they try to catch the ruler, or by having students close their eyes and rely on an aural cue to catch the ruler, like a whistle or on 'go'.

If there is enough time, collate results from individuals to get an average reaction time (per student) for the class, and compare this to published results for average human reaction time.

Review and practice

Stopping distances and motive forces (10 mins)

In this interactive, students link together parts of sentences on braking distance, identify true and false statements on stopping distance, reaction time, and braking distance, and order a description of forces involved as a car stops. Ensure students can use the proportionality between force and acceleration to appreciate the dangers of a large deceleration (e.g., the danger of a large, sudden braking force carried out in an emergency stop.)

Ⓗ P10.4 Momentum

AQA GCSE Physics specification points

Ⓗ 4.5.7.1 Momentum is a property of moving objects

Ⓗ 4.5.7.2 Conservation of momentum

Smart Curriculum Narrative

Forces and motion

Theme/subtheme

Forces and motion/Momentum

Disciplinary knowledge

- WS 1.2
- MS 3b, 3c

Practical

Investigating a controlled explosion

Learning objectives	Learning outcomes: Developing Aiming for 3	Learning outcomes: Secure Aiming for 5	Learning outcomes: Extending Aiming for 8
Recall and use the equation $p = mv$ to find the momentum of an object	Define momentum and describe it as a vector quantity	**Recall and use the equation $p = mv$ to find the momentum of an object**	Apply the equation $p = mv$ to find the momentum, velocity, or mass of an object
Describe how the principle of conservation of momentum can be used to predict the velocities of objects after a collision	State the principle of conservation of momentum	**Describe how the principle of conservation of momentum can be used to predict the velocities of objects after a collision**	Apply the principle of conservation of momentum to a range of situations to explain the behaviour of objects after a collision
Explain how your observations of explosions confirm the conservation of momentum	Observe the behaviour of objects during explosions	**Explain how your observations of explosions confirm the conservation of momentum**	Explain how frictional effects mean observed measurements are not the same as predicted data

Tier 2 vocabulary	Tier 3 vocabulary
energy, *mass*, scalar, speed, stationary, trolley, vector	*impact*, momentum, conservation of momentum, velocity

Lesson plan

Reactivate knowledge

1 What is the difference between a scalar and a vector quantity?

2 Car A has a velocity of 10 m/s and car B –10 m/s. Describe what you would see.

3 A force of 25 N acts upwards and force of 25 N acts downwards on an object. What is the resultant force?

Common misconception

Students may think momentum and kinetic energy are the same. Explain that momentum describes how much mass is in motion and in which direction (vector quantity), whereas the kinetic energy store simply describes how much energy an object has due to its motion (scalar quantity).

Answers: 1 Vector has size and direction, scalar has size *2* They travel at the same speed but in opposite directions *3* Zero

Trigger interest

Trying to stop (5 mins)

As a thought experiment, students should imagine a container ship (show images if helpful) and a cyclist, both moving at the same speed.

Students should consider which moving object will take longer and further to come to a stop, and their reasons why. Collect votes and a few suggestions, ensuring that ideas about both mass and speed (velocity) are explored.

To support, scaffold by introducing the converse scenario, for example, two cyclists of the same mass but moving at different speeds. This should help debate the significance of speed.

Exposition

Combining mass and velocity (15 mins)

Convince students that there must be an important property of moving objects to consider, which combines both mass and velocity. Introduce the equation for momentum, defining all symbols, and discussing examples. Emphasise that momentum is a vector, because velocity is a vector quantity.

The students should calculate the momentum of several moving objects, and the velocities of some objects when given mass and momentum. Support with an equation triangle, wherever needed.

Additional challenge can be introduced with examples using standard form, and/or SI prefixes.

Introduce the principle of conservation of momentum, the idea that, like energy, a conserved quantity means that the total amount before must equal the total amount after. Using a slow-motion video of a stationary, exploding object, students should state the momentum of the stationary object before the explosion as zero. Then, go through the clip frame-by-frame, discussing individual momenta of the exploding pieces to emphasise the principle of 'adding up the momenta', which must equal zero.

Main activity

Investigating a controlled explosion (25 mins)

Students use two trolleys, connected by a spring, to simulate a controlled explosion. They will use trial and error to place the blocks on the runway so that the trolleys reach them at the same time.

Students explain how it will demonstrate the conservation of momentum in more complicated explosions (with thousands of pieces). This will involve an explanation of the different velocities of the trolleys after the collision, and hence the distance they travel in the same time. Compare the data to their predictions and suggest explanations for any discrepancies.

Review and practice

Lost in space (10 mins)

Tell the students that an astronaut becomes tragically disconnected from her tether cable and drifts just out of reach of the spacecraft. She is unable to push or pull on anything in order to return. Luckily, she has a small fire extinguisher that she uses to return to safety.

Students discuss how the astronaut used the fire extinguisher to return to the spacecraft. Where necessary, use a sketch to scaffold students' ideas by asking, for instance, how some momentum to the right might be generated by giving something else (e.g., an exhaust fume or the fire extinguisher itself) some momentum in another direction. Link the idea that momentum must be conserved, so zero momentum before should equal zero momentum after, because the objects will give each other the same equal and opposite momentum.

Ⓗ P10.5 Using conservation of momentum

AQA GCSE Physics specification points

Ⓗ 4.5.7.2 Conservation of momentum

Smart Curriculum Narrative

Forces and motion

Theme/subtheme

Forces and motion/Momentum

Disciplinary knowledge

- WS 1.2
- MS 3b, 3c

Practical

Investigating and analysing collisions

	Learning outcomes		
Learning objectives	**Developing Aiming for 3**	**Secure Aiming for 5**	**Extending Aiming for 8**
Apply the law of conservation of momentum to find velocities of objects after impacts	Determine the total momentum before or after impacts	**Apply the law of conservation of momentum to find velocities of objects after impacts**	Apply the law of conservation of momentum to many different scenarios
Determine the momentum before and after collision using results	Measure mass and velocity of objects before a collision	**Determine the momentum before and after collision using results**	Determine the percentage difference between expected and actual data collected
Evaluate data collected to verify the law of conservation of momentum	Identify sources of error in measurements	**Evaluate data collected to verify the law of conservation of momentum**	Evaluate an experimental technique and discuss in detail the factors that lead to differences between experimental data and an accepted law

Tier 2 vocabulary	Tier 3 vocabulary
frictional force, resultant force, scalar, *vector*	*impact*, kinetic energy, conservation of momentum, velocity

Lesson plan

Reactivate knowledge

1 What formula to you use to calculate momentum?

2 State the conservation of momentum.

3 Is momentum a vector or scalar quantity?

Answers: 1 $p = mv$ *2* Total momentum before an event equals total momentum after the event in a closed system *3* Vector quantity

Common misconception

Students may think momentum and kinetic energy are the same. Explain that there are similarities between momentum and kinetic energy. Both are properties of moving objects. However, momentum describes how much mass is in motion and in which direction (vector quantity), whereas the kinetic energy store simply describes how much energy an object has due to its motion (scalar quantity).

Trigger interest

Balloons conserve momentum (5 mins)

Inflate a balloon without tying it. Ask students to predict and describe the motion of the balloon after it is let loose. If appropriate, provide balloons to pairs of students and encourage them to carry out the process themselves, presenting descriptions with diagrams on mini-whiteboards.

Support students by providing keywords to use in their answers, for example, students may use some or all of the following: accelerate, resultant force, conservation of momentum, initial momentum, final momentum, opposite direction. The purpose of this demonstration is to lead students to the idea that forward (or upward) momentum is balanced by backward (or downward) momentum, if the total momentum before the event is zero.

Extend this discussion by asking students to compare and contrast the motion of a rising ballon with a rocket, which is another demonstration of the conservation of momentum.

Exposition

Scenarios that conserve momentum (15 mins)

Students should analyse scenarios where trolleys move apart, performing calculations to find the velocities of the components immediately after 'explosion'. Students can answer the end of spread questions in the Student Book for practice, before moving on to the practical investigation.

Support students by displaying a calculation framework that leads them through the stages of the calculation. It will help, for instance, to be very explicit with equating momentum *before* to momentum *after*, keeping all corresponding working in columns under these headers.

Additional challenge is provided by allowing students to find the distance travelled by the components a specific time after the explosion.

Main activity

Investigating and analysing collisions (25 mins)

Students analyse collisions using the practical. Ensure that students are calculating the momentum before and after the impacts, and then comparing them, with the expectation that they are equal.

Challenge students, where appropriate, to also describe the changes in energy stores of the system, explaining these changes (and any discrepancies in the conservation of momentum) in terms of the frictional forces acting.

Review and practice

Impossibly super (10 mins)

In many superhero films, cars and other large objects are seemingly thrown and caught with ease. Students use the interactive activity to complete an explanation about the flaws in these films with respect to the conservation of momentum. They can then consider how different forces would impact on the system (do they think that the friction of the feet on the ground would make any significant difference?)

Ⓗ P10.6 Impact forces

AQA GCSE Physics specification points

Ⓗ 4.5.7.3 Changes in momentum

Smart Curriculum Narrative

Forces and motion

Theme/subtheme

Forces and motion/Momentum

Disciplinary knowledge

- WS 1.2
- MS 3a, 3b, 3c

Practical

Investigating impacts

Learning objectives	Learning outcomes: Developing Aiming for 3	Learning outcomes: Secure Aiming for 5	Learning outcomes: Extending Aiming for 8
Describe how the impact force changes when the change in momentum occurs over an increased impact time	State that a force acting on a moving object causes a change in momentum	**Describe how the impact force changes when the change in momentum occurs over an increased impact time**	Explain how the impact force changes if the impact time is halved
Calculate the impact force using the change in momentum and the time of impact	State the factors affecting the size of the impact force	**Calculate the impact force using the change in momentum and the time of impact**	Calculate changes in velocity and momentum during impacts using the force involved in the impact and the impact time
Show how the equations for resultant force and acceleration combine to give the relationship impact force is equal to rate of change of momentum	State Newton's Second Law	**Show how the equations for resultant force and acceleration combine to give the relationship impact force is equal to rate of change of momentum**	Apply the concept of equal and opposite forces in collisions to explain why momentum is conserved in impacts

Tier 2 vocabulary	Tier 3 vocabulary
force, inversely proportional, *mass*, scalar, *vector*	impact force, kinetic energy, velocity

Lesson plan

Reactivate knowledge

1 State Newton's Second Law.

2 How do you calculate acceleration?

3 How do you calculate momentum?

Answers: 1 $F = ma$ 2 acceleration = change in velocity/time 3 $p = mv$

Common misconception

Students may think momentum and kinetic energy are the same. Explain that there are similarities between momentum and kinetic energy. Both are properties of moving objects. However, momentum describes how much mass is in motion and in which direction (vector quantity), whereas the kinetic energy store simply describes how much energy an object has due to its motion (scalar quantity).

Trigger interest

An egg-cellent demo (10 mins)

Two student volunteers hold a large linen sheet, or thin cloth, outspread but not too tight. Drop an egg from a height of one or two metres and it will not break. If possible, throw the same egg at the sheet (a hard throw adds to the dramatic effect!) and again it does not break.

Students should discuss, in pairs, what this demo tells them about impact forces and contact time, given the fact that the same egg if dropped/thrown against a hard surface will shatter. Allow students to formulate some opinions and encourage them to see if the upcoming lesson content reaffirms or changes their conclusions.

Support students where needed with a sketch of the impact force on the egg from the surface of the cloth (and/or hard surface), and a verbal explanation of 'impact time' as the time during which the impact force acts on the egg.

Exposition

Impact force and time (10 mins)

Introduce students to the central concept that when a force acts on an object, it causes a change in momentum. Link this to the relationship $F = \frac{(m\Delta v)}{\Delta t}$ where $m\Delta v$ is the change in momentum. Ask students to describe what this equation tells us about the relationship between the impact force F and the impact time Δt, drawing out any mathematical recognition of inverse proportionality. It may be necessary, for instance, to show a few quick examples of how halving one variable doubles the other, and quartering one quadruples the other, etc. They should understand that a longer impact time reduces the impact force for a given change in momentum, for example, to bring an object to rest, and vice versa.

Support students with a couple of examples of impact forces and times, for example, a video clip showing the crumple zone on a car, and crash dummies testing car seatbelts. It is helpful, wherever possible, to sketch the vector of the force acting on the object in question, so that students can visualise the size and direction of this force and how long it is in contact with the object.

Extension opportunities involve descriptions of the pair of forces acting using Newton's Third Law, which says that the object exerts an equal and opposite force on the surface it collides with.

Main activity

Investigating impacts (30 mins)

Students investigate the effect of crumple zones or similar using the simple practical. The calculations required are quite complex and must be broken down into logical stages. As you circulate pairs or groups, remind students of some simple acceleration calculations, highlighting deceleration examples (final velocity is zero) as this is more important in impacts. Repeat the definition and calculation of the quantity $m\Delta v$ as the change in momentum, before linking this to the size of the acting force.

Review and practice

An egg-cellent demo (revisited) (5 mins)

Create five short multiple-choice questions analysing the impact force and impact time on the egg, giving students a short time to think and vote with mini-whiteboards. Questions should gently increase in difficulty, assessing students' understanding of the final velocity of the egg (0 m/s), the quantity, the impact time, and an explanation of why it does not break.

Ⓗ P10.7 Safety first

AQA GCSE Physics specification points

Ⓗ 4.5.7.3 Changes in momentum

4.5.6.3.1 Stopping distance

Smart Curriculum Narrative

Forces and motion

Theme/subtheme

Forces and motion/Momentum

Disciplinary knowledge

- WS 1.2, 1.4
- MS 2c, 2f, 3b, 3c, 3d

Activity

Hard sell

Learning objectives	Learning outcomes: Developing Aiming for 3	Learning outcomes: Secure Aiming for 5	Learning outcomes: Extending Aiming for 8
Explain the operation of some safety features of a car in simple terms	State some examples of car safety features	**Explain the operation of some safety features of a car in simple terms**	Explain using rate of change of momentum how air bags reduce injuries
Explain how one of these safety features reduces impact force	Give examples of other safety features such as cushioned surfaces in playgrounds	**Explain how one of these safety features reduces impact force**	Evaluate a range of optional safety features based on their costs and effectiveness

Tier 2 vocabulary	Tier 3 vocabulary
elastic, energy, *plastic*, speed	absorb, collision, crumpled, impact force, impact time

Lesson plan

Reactivate knowledge

1 How does impact time effect impact force?

2 How do you calculate impact force?

3 What is inertia?

Answers: 1 Increasing the impact time decreases the impact force *2 F = ma 3* The tendency of an object to stay at rest or keep moving with constant velocity

Common misconception

Students may think that people who drive in a city have greater need for car safety features than those who drive in rural areas. Respond by saying car safety features can help protect drivers in all kinds of climates, road and traffic conditions, and terrains. Although there are different challenges and hazards associated with city and rural driving, car safety features are always essential protection mechanisms. For example, in a city, there may be more traffic and people which create risk, but in a rural area, visibility may be reduced due to winding or narrow roads, and road conditions could be more variable.

Trigger interest

Crumpled cars (10 mins)

Students use the interactive to identify safety features of a car and the reasons why a car engineer would specifically design these features.

Support can be provided by refreshing students' ideas, linking to general properties about plastic and elastic materials, for example, flexibility, malleability, and stretchiness. Additional conversations can explore the energy changes involved in collisions.

Exposition

Car safety features (introduction) (15 mins)

Show the students a cycle helmet. In pairs, they should discuss and list as many design features (shape, materials, colour, etc) that are used to make this a protective piece of headgear. As you circulate, introduce the idea of 'cushioning' an impact and ask them to think about what this might mean, revisiting last lesson's use of the terms 'impact time' and 'impact force'.

Next, provide pairs of students with images of different car safety features (including seat belts, child car seats, and air bags), allowing time for discussion and ideation about their design principles.

Ensure students understand that soft materials absorb the energy from impacts (rather than human beings!), while strong materials maintain the shape and structure (e.g., of a child car seat) during collision.

Main activity

Hard sell (25 mins)

Provide students with some historical context about the safety features of cars, for example, that most safety features started off as expensive accessories but as their benefits were proven, they became required by law.

Students must then produce an advertisement for expensive safety features in a new car. This could be hard-hitting, or subtler in its persuasiveness. Encourage students to be creative with the resources available, for example, they could create a short video, or digital poster.

Include a requirement that the science behind the safety feature(s) must be clear, which will make their sell more likely, that is, information about impact force and time, materials, energy absorption, and so on.

If there is time, provide students a simple framework for peer-assessment. For example, students could assess each other's advertisements based on a short set of criteria, and vote on which is most persuasive for the strongest 'buy'.

Review and practice

Best car (5 mins)

Show students a slideshow of cars through the ages and get them to identify the innovations that have taken place with respect to road safety. Include broader ideas about changes to maximum speeds and road types, and the introduction of motorways.

A forward look to new innovations, like self-driving technology, would end the lesson well.

P10.8 Forces and elasticity

AQA GCSE Physics specification points

4.5.3 Forces and elasticity

Smart Curriculum Narrative

Forces and motion

Theme/subtheme

Forces and motion/Deformation and Hooke's Law

Disciplinary knowledge

- MS 3b, 3c, 4a, 4b

Required Practical 6

Force and extension

Learning objectives	Learning outcomes: Developing (Aiming for 3)	Learning outcomes: Secure (Aiming for 5)	Learning outcomes: Extending (Aiming for 8)
Compare materials in terms of elastic and non-elastic behaviour	Explain why more than one force is needed to change the shape of an object	**Compare materials in terms of elastic and non-elastic behaviour**	Compare the behaviour of different materials under loads in terms of linear and non-linear behaviour
Calculate the force required to cause a given extension (or compression) in a spring using the spring constant	State Hooke's law	**Calculate the force required to cause a given extension (or compression) in a spring using the spring constant**	Apply the equation $F = ke$ in a wide range of situations
Use the gradient of a force extension graph to determine the spring constant of a spring	Investigate the relationship between force and extension for a spring	**Use the gradient of a force extension graph to determine the spring constant of a spring**	Evaluate an investigation into the extension of materials in terms of the accuracy of your measurements and the resolution of your measuring instruments

Tier 2 vocabulary	Tier 3 vocabulary
clamp, directly proportional, force, gradient, polythene, *spring*	compression, distortion, extension, force–extension graph, spring constant, stretch

Lesson plan

Reactivate knowledge

1 What is an elastic material?

2 Give an example of an inelastic material.

3 The spring constant for a material is 25 N/m, what does this mean?

Common misconception

Students may think the extension of a spring or material is the new length. Tell students how much a spring or material extends (or compressed) can be calculated by comparing the original length to the new extended (or compressed) length.

Answers: 1 One that returns to its original shape when the force is removed *2* Example: sticky tac/putty *3* A force of 25 N will extend the spring by 1 m/it shows how easy it is to stretch or compress the material

Trigger interest

Distortion (5 mins)

Forces can change an object's speed, direction, or shape. Create a sorting activity for groups of students to identify the effects of these forces. A collection of images for each group will work well; alternatively, use a series of video clips and students can vote between one of the three options.

Exposition

That's a stretch (10 mins)

Ask student volunteers to demonstrate applying a force to simple objects, showing them stretching and returning to their original shape. For example, simple elastic bands, an ordinary spring, and a large slinky spring. If available, show the plastic behaviour of polythene or something similar, whilst also stretching a spring beyond its elastic limit, asking students to describe what has happened.

Introduce the equation $F = ke$ defining each symbol while setting up a demonstration using a mass hanger, retort stand, clamp, spring, and metre rule. Show students how you determine the extension, emphasising the distinction between original and final length. Show students explicitly how to calculate the force from the given mass using $W = mg$ (although if needed, provide these values in advance). For the practical below, it can be assumed that $50\,g = 0.5\,N$, $100\,g = 1\,N$, etc, as this is sufficient for the learning purposes and can avoid calculations giving hard-to-use values.

Give one example to calculate the spring constant using measured values of F and e, introducing the spring constant that describes how much force is required to extend a spring (per unit of length).

Ensure that students understand the purpose of the practical is to investigate the relationship between F and e, to find out whether they truly are proportional to each other, as the equation suggests.

Main activity

Force and extension (30 mins)

Students use the practical to test the behaviour of materials under load. They should focus on accurate measurement of length and extension, and well-organised recording of data. It is important to emphasise that the plotting of a graph is not the end in itself, but rather, the graph is a tool for discovering or confirming any relationship between variables – in this case, force and extension.

Identify areas on students' graphs where the relationship is directly proportional, ensuring students can identify this. Explore ideas about where limits to this behaviour are observed, linking back to the starter activity. Students should find the spring constant using the gradient of the graph, because $k = \frac{F}{e}$

For additional support, circulate the room with a sketch of the graph, or with your own results and graph, to enable analysis and discussion of the key concepts if students struggle to complete their own measurements in time.

Review and practice

Graphical analysis (10 mins)

Show students three or four different sketches of force–extension graphs and ask them to identify the key components from these sketches. For example, direct proportionality and bigger/smaller spring constants from the gradient.

P10 Forces and motion

Introduction to checkpoint intervention

The checkpoint intervention provides suggestions for a lesson to follow up the *P10 Forces and motion* checkpoint assessment. Depending on the outcomes of the assessment, these suggestions can be used to move a class towards secure, to target specific learning outcomes, or to consolidate knowledge for students achieving secure outcomes. Use students' outcomes from the checkpoint assessment to address general misconceptions from the content covered. Be prepared to re-cover content as required; three key concepts from the chapter are covered in more detail below.

Checkpoint secure learning outcomes

- Apply the concept of balanced forces to explain why an object falling through a fluid will reach a terminal velocity.
- Determine the thinking distance using reaction time and initial velocity.
- Use and rearrange the $F = ma$ equation to find the resultant force acting on an object, its mass, or its acceleration.
- Apply the law of conservation of momentum to find velocities of objects after impacts.
- Describe how the impact force changes when the change in momentum occurs over an increased impact time.
- Compare materials in terms of elastic and non-elastic behaviour.

Student reflection

Encourage students to reflect on whether there were any checkpoint questions they found difficult or straightforward as well as their preparation for the checkpoint assessment, including how they revised, what they revised, the time spent on revision, and what areas they could improve upon. For more strategies, see the Metacognition spread.

Intervention activity: Apply the concept of balanced forces to explain why an object falling through a fluid will reach a terminal velocity.

The specialness of constant velocity

Give students plenty of examples (images, video clips, and animations) of simple objects moving at constant velocity, that is, constant speed in a straight line, and stationary objects.

Begin this intervention by encouraging students to identify motion with constant velocity, emphasising the 'specialness' of this state of motion. It can be thought of as special because it only occurs in very specific circumstances. Once they are comfortable with this foundational idea, gradually introduce notions of force.

Balanced forces and terminal velocity

Focus on scenarios in which two unbalanced forces cause objects to change speed. Ensure these forces are labelled – type and direction – so that students can focus on the effect of them being unbalanced.

Once students can identify cases in which an object's speed is changing, given two opposing forces acting, introduce the idea of terminal velocity by showing a sequential series of images for a falling skydiver. For instance, the first still image could be just as they leave the plane: label a large downward vector for the weight, and a small upward vector for air resistance.

The more images in this sequence, the better. All but one image should illustrate the skydiver's changing speed, since weight does not equal air resistance until terminal velocity is reached. Explain why this happens gradually – the final key concept that students should recognise is that the weight stays constant while air resistance increases as the skydiver falls – hence, these forces will eventually balance each other.

Intervention activity: Use and rearrange the $F = ma$ equation to find the resultant force acting on an object, its mass, or its acceleration.

Defining the terms

Ensure that students can define and explain each of the terms in the equation $F = ma$

Start with mass, m. It is fine if students express this as a measure of how much matter an object contains, and even better if they agree that more matter is more difficult to start in motion (or, conversely, to stop motion) Next, illustrate simple cases of acceleration, a, representing an object's change in speed. Students should appreciate that m/s^2 means the speed changes every second.

Finally, link a resultant force to the acceleration, such that students can clearly visualise the idea that a (resultant) force causes objects to speed up or slow down. The first part of this intervention should focus on discussion, using visual aids and sketches wherever possible, to fully conceptualise these quantities before using them mathematically.

Using $F = ma$

Ask students to predict the relationship between resultant force F and the acceleration, a, caused for the same object (i.e., for constant mass). Encourage them to use their intuition to see that these are proportional, before showing a calculation using simple integers, for example, the force required for an object of mass 10 kg to accelerate at $2\,m/s^2$ and then to accelerate at $4\,m/s^2$.

Show how to substitute numbers into the equation to get an answer for F explicitly, before moving on to more difficult examples requiring rearranging to calculate m or a. Wherever needed, use an equation triangle or other framework to support rearranging.

Intervention activity: Describe how the impact force changes when the change in momentum occurs over an increased impact time.

Bringing objects to rest

A useful scenario to illustrate this objective is collisions that bring an object to rest. Ensure that students can identify these (i.e., cases for which the final velocity is zero) and that they can calculate the corresponding change in momentum, given an initial velocity. Video clips of Newton's cradle, or snooker balls being hit such that the incoming ball is being brought to a sudden standstill, work well.

Impact force and impact time

Ask students to recall the equation that links changing momentum with force. Ensure they can verbalise this equation (i.e., "force is equal to the rate of change of momentum") and that what this actually means is understood. For instance, the size of a force on an object determines how *quickly* its momentum changes.

Students should then identify the mathematical relationship between impact force and impact time. Demonstrate, using a falling egg on a taut sheet or thin cloth, that the impact time is very long because the egg is in contact with the sheet for a long time during its collision. This means the impact force on the egg is reduced and the sheet does not break the egg in bringing it to rest (unlike a hard surface). Once this is understood, students should explore the various ways in which impact forces from collisions are reduced by increasing impact time in the context of car safety. For example, the design of a crumple zone, seat belts, and airbags. Other examples can be explored by demonstration, for example, 'catching' an object where someone lengthens the impact time using a catching movement, or padded gloves.

Physics only

P11 Force and pressure

P11.1 Pressure and surfaces

AQA GCSE Physics specification points
4.5.5.1.1 Pressure in a fluid 1

Smart Curriculum Narrative
Forces and motion

Theme/subtheme
Forces and motion/Pressure

Disciplinary knowledge
- WS 4.3, 4.4, 4.5, 4.6
- MS 3b, 3c

Practical
Measure your hand pressure

	Learning outcomes		
Learning objectives	**Developing Aiming for 3**	**Secure Aiming for 5**	**Extending Aiming for 8**
Describe the effect on pressure of changing the area of contact or weight acting on a surface	State the factors that affect the pressure acting on a surface	**Describe the effect on pressure of changing the area of contact or weight acting on a surface**	Apply the concept of pressure in explaining the effect on a surface in a wide range of contexts
Use the equation for pressure to determine pressure, area, or force	Calculate the pressure caused by an object resting on a surface, given the force and area of contact	**Use the equation for pressure to determine pressure, area, or force**	Perform pressure calculations including conversion of areas and forces with SI multiplier prefixes
Describe how pressure in fluids causes a force perpendicular to the surface	State that fluids (liquids or gases) exert a pressure on surfaces	**Describe how pressure in fluids causes a force perpendicular to the surface**	Determine the force acting on an area when given the pressure of a fluid

Tier 2 vocabulary	Tier 3 vocabulary
area, force, Newton, proportional	pascal, pressure

Lesson plan

Reactivate knowledge

1 Give an example of high and low pressure acting on a surface.
2 What is the difference between force and pressure?
3 How can you increase the pressure acting on a surface?

Common misconception

Students may think pressure and force are synonymous. Tell students that force is the push or pull that causes objects to change velocity. Pressure is a measure of how a force is spread out, or how it is concentrated, over a certain area.

Answers: 1 High: knives, nails, thumb tacks, holding a string handle; Low: snowshoes, skis, caterpillar tracks *2* Pressure is force per unit area *3* Increase the force and/or decrease the area

Trigger interest

Let that sink in (10 mins)

Provide a large tray of damp sand and a variety of objects – some with larger flat surfaces, and some more pointed or spikey. Ask students to predict what will happen as they press the objects into the sand. For example, cylindrical objects compared to similar-sized cones work well for this demonstration; students should appreciate the ease with which the tip of a cone sinks into the sand compared to the flat surfaces of cylinders. Ask for explanations of why some sink more easily than others.

If the tray is large enough, students could step into it wearing different types of shoes, for example, some with heels and some with a large flat surface.

Extend this discussion by asking students to explain why the objects eventually stop sinking into the sand (assuming it has not reached the bottom).

Exposition

Conceptualising pressure (15 mins)

Use the conclusions of the starter to discuss the concept of pressure, using plenty of visual examples such as those shown in the Student Book. Students should become familiar with the action of a force acting over an area.

Introduce the equation relating pressure to force and area. Use several examples, including rearrangement, some of which can be taken from the summary questions in the Student Book. It is useful to emphasise the proportionality of pressure to force, and its inverse proportionality to area. Similarly, emphasise the unit for pressure (pascal) and its equivalence to the Newton per square metre – use these terms interchangeably but remind students that the standard international unit is the pascal.

Main activity

Measure your hand pressure (20 mins)

Use of the pressure equation can be reinforced by asking students to individually determine the maximum pressure they can exert with a hand. Students push on a set of weighing scales, using the value to determine their maximum exerted force. They draw around their outstretched hand on cm squared paper, determining the area, and calculate their maximum exerted pressure.

Support students wherever needed to determine the force from the reading on the weighing scales. Students will likely need support converting from cm^2 to m^2, if the answer is needed in Pa; alternatively, answers can be left in N/cm^2. The process of counting cm squares to approximate an area is important; ensure students are comfortable showing how they carry out this process, for example, by shading individual squares and approximating fractional squares.

Review and practice

On thin ice (10 mins)

Show students video clips of polar bears crossing thin ice by lying and spreading their bodies out. This is an interesting illustration of the pressure equation applied instinctively by animals in nature.

Students then complete the interactive activity, identifying measures that must be taken during a rescue on an ice-covered lake, and discuss the measures taken to reduce the pressure on the ice.

Support can be introduced here by showing additional methods of spreading force over large areas to reduce pressure and the chances of sinking, for example, snowshoes and snowboards.

Ⓗ P11.2 Pressure in a liquid at rest

AQA GCSE Physics specification points

Ⓗ 4.5.5.1.2 Pressure in a fluid 2

Smart Curriculum Narrative

Forces and motion

Theme/subtheme

Forces and motion/Pressure

Disciplinary knowledge

- WS 4.3, 4.4, 4.5, 4.6
- MS 3b, 3c

Activity

Pressure calculations

Learning objectives	Learning outcomes: Developing – Aiming for 3	Learning outcomes: Secure – Aiming for 5	Learning outcomes: Extending – Aiming for 8
Explain why liquid pressure increases with depth and density	State that pressure in a liquid changes with depth and density	**Explain why liquid pressure increases with depth and density**	Using pressure equation, density equation and $W = mg$, use algebraic techniques to derive the equation $p = h\rho g$
Calculate the pressure at a point in a liquid using $p = h\rho g$	Describe how liquid pressure changes with depth and density	**Calculate the pressure at a point in a liquid using $p = h\rho g$**	Rearrange the equation $p = h\rho g$ to solve a range of questions involving the pressure in a liquid

Tier 2 vocabulary	Tier 3 vocabulary
height, proportional, weight	density, pressure

Lesson plan

Reactivate knowledge

1 How do you calculate pressure?
2 How do fluids exert a pressure on a surface?
3 What is density?

Common misconception

Students may think that liquid pressure is the same at all points in a liquid. Respond that liquid pressure increases with depth (also referred to as height) from the surface, which means objects deeper in a liquid experience a greater liquid pressure. This is because the deeper an object is under liquid, there is more liquid above it – so there is a larger weight force due to that liquid acting on the area of the object. This results in a larger liquid pressure.

Answers: *1* Force ÷ area *2* Particles colliding with the surface exerting a force over an area *3* Density is the mass per unit volume

Trigger interest

The deep (10 mins)

Students use the interactive activity to describe the conditions deep beneath the oceans and how this relates to the design of submarines. They should realise that the pressure is very large even at fairly shallow depths – in fact, pressure doubles at only 10 m below the surface.

Where appropriate, encourage students to carry out independent research to explore how deep-sea creatures adapt to such high-pressure environments. Blobfish are great case studies of biophysics in action.

Exposition

Pressure tests (20 mins)

Explain the factors that affect the pressure at a specific depth in a liquid using the three simple demonstrations described in the practical. These effects need to be explained by discussing the forces acting at specific points – forces provided by the weight of the fluid above the holes in the bottle and the pipe. This will lead onto discussion of the calculations.

Link the equations for weight, $W = mg$, density, $\rho = \frac{m}{V}$, and pressure, $p = \frac{F}{A}$ to derive the equation for liquid pressure at height, h (as shown in the Student Book).

Ensure students are clear about each of the terms in the equation. For example, they should be able to sketch and label the distance, h, from the surface of the sea to a blobfish.

Reinforce students' intuition with the proportionalities in the equation by asking them to state how the liquid pressure changes if density or height changes. They should be able to make statements such as: 'If the density of the liquid increases, the pressure increases'.

Main activity

Pressure calculations (20 mins)

Students should try plenty of calculations applying the liquid pressure equation. Provide a calculation support frame where needed, especially in questions that require rearrangement.

Review and practice

Pascal's vases (5 mins)

Ask students to discuss and suggest, in pairs, their predictions for where the fluid will settle as it is poured into this equipment. They should justify their predictions, linking them to the idea of equal pressure at equal depth, as shown in the practical demonstration from earlier in the lesson.

P11.3 Atmospheric pressure

AQA GCSE Physics specification points

4.5.5.2 Atmospheric pressure

H 4.5.5.1.2 Pressure in a fluid 2

Smart Curriculum Narrative

Forces and motion

Theme/subtheme

Forces and motion/Pressure

Disciplinary knowledge

- WS 1.2, 4.3, 4.4, 4.5, 4.6
- MS 3b, 3c

Practical

Atmospheric pressure and its effects

	Learning outcomes		
Learning objectives	**Developing Aiming for 3**	**Secure Aiming for 5**	**Extending Aiming for 8**
Describe what causes atmospheric pressure	State that the air around you exerts a pressure	**Describe what causes atmospheric pressure**	Calculate the force on a flat object due to a pressure difference
Describe a simple model of the Earth's atmosphere	State that the Earth's atmosphere is a thin layer of air surrounding the Earth	**Describe a simple model of the Earth's atmosphere**	Compare the model of the Earth's atmosphere to real data
Explain why atmospheric pressure varies with height	State that the density of the atmosphere decreases with height	**Explain why atmospheric pressure varies with height**	Explain why the relationship $p = h\rho g$ is not suitable for calculating changes in pressure in the atmosphere over a large change in height

Tier 2 vocabulary	Tier 3 vocabulary
area, atmosphere, equilibrium, height, magnitude	density, gravitational field strength, pressure

Lesson plan

Reactivate knowledge

1 What is the air around you made up of?

2 Describe the particle model for a gas.

3 What is the difference between the number of particles for a gas at high density and a gas at low density?

Common misconception

Students may think that the atmosphere doesn't exert any pressure because we don't feel it. Explain it as a vast gas ocean overhead, composed of various gases with particles. Collectively, these particles weigh significantly, exerting pressure on Earth's surface. This atmospheric weight, approximately 100,000 Pa, is what we experience as atmospheric pressure. Life on Earth evolved to tolerate this pressure, with our bodies' outward pressure typically in balance with it, hence why we don't sense it.

Answers: 1 Mixture of gases/(79%) nitrogen, (21%) oxygen, and other gases *2* Particles spaced apart and move in random directions *3* The one at high density will have more particles per unit volume

Trigger interest

Atmospheric pressure (10 mins)

Find out what students already know about atmospheric pressure. Ask if they have experienced their ears popping when they change altitude quickly (e.g., going up or down in a very fast elevator, driving up a steep hill/mountain, or flying in an airplane). Some may link atmospheric pressure and weather patterns.

Explain the importance of the atmosphere to all life on Earth. An image of the atmosphere is a nice visual prompt to emphasise how small it is in comparison to the radius of the Earth, and how fortunate we are to have one rich in the elements needed to sustain life on the Earth's surface. As a prelude to the interactive, show the 'can-crushing' demo to demonstrate the reality of atmospheric pressure; then get students to use the interactive to describe its causes.

Exposition

The weight of the sky (15 mins)

Link back to P11.2 and the factors that affect pressure in a liquid – namely, the density of the liquid, the gravitational field strength, and the height within the liquid. Since this lesson is about gases, ask students to describe the difference between liquids and gases, in terms of density and the motion of particles.

Using an animation or simulation of gases, demonstrate how gas pressure is exerted on surfaces. It is helpful to 'zoom in' on the motion of a single gas particle, and the collision against a surface exerting a small force. Students should be able to appreciate that, for an atmosphere of trillions of small particles exerting forces constantly, the sum is a very large force! Link this to pressure by considering a square metre, emphasising that the type of surface does not matter (i.e., atmospheric pressure is constant at ground level).

Extend this discussion by exploring the effect of reduced temperature on the motion of gas particles and the resulting effect on atmospheric pressure.

Main activity

Atmospheric pressure and its effects (20 mins)

Demonstrate the effect of a suction cup by attaching it to a small metal plate, or stool, and lifting it. If Magdeburg hemispheres are available, this is a nice alternative; demonstrate the effect of removing air from them and then asking student pairs to attempt to pull them apart.

Ensure students can describe the effect of removing air, in terms of the reduction of collisions (forces) against the interior surface of the hemispheres. A sketch of the resulting effect, and the overwhelming influence of the surrounding atmosphere, would be useful.

Higher tier students should relate atmospheric pressure to the equation $p = \rho g h$ and discuss why this is an appropriate equation to determine the atmospheric pressure at any height.

All students should compare the equation with atmospheric pressure graphs and note that they do not match. The equation would produce a linear relationship between p and h; students should discuss the assumptions about density in the atmosphere vs the reality, to account for this difference.

Review and practice

Suction? (10 mins)

Stick a suction cup toy on a window and ask students to discuss whether 'suction' is a misleading word to describe the effect observed. Ensure answers touch on the effect of atmospheric pressure, in terms of forces on the outer surface of the toy.

A diagram is useful for additional support here to reinforce the idea that atmospheric pressure (large forces acting on each unit of area) acts externally on the suction cup.

Ⓗ P11.4 Upthrust and flotation

AQA GCSE Physics specification points

Ⓗ 4.5.5.1.2 Pressure in a fluid 2

Smart Curriculum Narrative

Forces and motion

Theme/subtheme

Forces and motion/Types of force

Disciplinary knowledge

- MS 1c, 3c

Practical

Investigating upthrust

	Learning outcomes		
Learning objectives	Developing Aiming for 3	Secure Aiming for 5	Extending Aiming for 8
Explain why there is an upwards force	State that upthrust is the upwards force acting on an object in a fluid	**Explain why there is an upwards force**	Use algebraic techniques to show that the weight of liquid displaced is equal to the upthrust provided
Describe the relationship between upthrust and weight for floating and submerged objects	Describe the forces acting on an object in a fluid	**Describe the relationship between upthrust and weight for floating and submerged objects**	Calculate the upthrust acting on a submerged object by using the pressure difference on the top and bottom surfaces
Investigate the factors that affect whether an object will float or sink	Describe the factors that affect whether an object will float or sink	**Investigate the factors that affect whether an object will float or sink**	Carry out and evaluate in detail an investigation into the factors that affect whether an object floats or sinks

Tier 2 vocabulary	Tier 3 vocabulary
cylinder, force, mass, resultant force, weight	density, Newton meter, pressure, upthrust

Lesson plan

Reactivate knowledge

1 Name the forces acting on a floating object.

2 How do these forces compare?

3 What does the pressure in a liquid depend on?

Common misconception

Students may think objects float on water, or a liquid, because they are lighter. Tell students that floating objects are stationary, so they have to have balanced forces acting on them. The weight of the object must be balanced by an upwards force acting on the object, which is called the upthrust. It can be shown that the density of the object, rather than its mass, determines whether it sinks or floats in water – generally, if an object is denser than water, it will sink.

Answers: 1 Weight and upthrust 2 They are equal 3 Depth and density

Trigger interest

Float or sink? (10 mins)

Use a large tank (e.g., a fish tank or large plastic container) to test whether a range of objects will float or sink. Students should predict, using mini-whiteboards, the outcome for each object.

Gather ideas about the factors that influence sinking or floating. At this stage, it's fine to collect broad responses – praise the use of keywords like density, pressure, and force, and write them down on a white-board or poster to refer to throughout the lesson.

Exposition

An upwards force (15 mins)

Demonstrate a floating object and ask students to discuss how many forces must be acting on the floating object. They should sketch their ideas as a free-body diagram on mini-whiteboards. Ensure they are convinced that an upwards force must exist on any floating object, to balance its weight in the fluid.

Introduce the keyword upthrust, adding this to the collection of keywords collected in the starter activity. The cause of upthrust should be explained with reference to the equation for pressure at a depth in a liquid – show a diagram of a large, submerged cylinder, leading students to conclude that the liquid pressure at the bottom surface must be greater than the pressure at the top surface. Therefore, this difference in pressure (or forces on each surface) require a resultant force upwards, which is upthrust.

Show images of large ocean liners, for example, cruise ships and oil tankers. Allow students to imagine the large upthrust forces required, and provide opportunity to demonstrate how these seemingly complex floating objects are simplified in terms of the two forces acting on them.

Main activity

Investigating upthrust (25 mins)

Frame this practical as an exploration of the effect of upthrust on a submerged (or partially submerged) object. Students lower objects into water and note the apparent reduction in weight using a Newton meter.

Circulate groups and discuss why some objects tend to sink or float in terms of comparing the weight and upthrust. Carry a simple sketch showing each of the two scenarios – equal and unbalanced forces on the floating or sinking object – which should help reinforce the idea that the size of the upthrust determines the fate of objects in a liquid.

Extend this idea by exploring the proof that upthrust is equal to the weight of the fluid displaced using the activity. Frame this as an explanation as to why some objects can 'generate' a larger upthrust force than others. A real-life example of floating objects displacing large volumes of liquid, for example, an oil tanker, is a useful illustration of this consequence of upthrust.

Review and practice

The pressure is off at last (5 mins)

Use the interactive to give students a series of true or false statements covering the content from the whole chapter. Encourage students to correct false statements.

P11 Force and pressure

Introduction to checkpoint intervention

The checkpoint intervention provides suggestions for a lesson to follow up the *P11 Pressure and surfaces* checkpoint assessment. Depending on the outcomes of the assessment, these suggestions can be used to move a class towards secure, to target specific learning outcomes, or to consolidate knowledge for students achieving secure outcomes. Use students' outcomes from the checkpoint assessment to address general misconceptions from the content covered. Be prepared to re-cover content as required; three key concepts from the chapter are covered in more detail below.

Checkpoint secure learning outcomes

- Use the equation for pressure to determine pressure, area, or force.
- Describe how pressure in fluids causes a force perpendicular to the surface.
- Explain why liquid pressure increases with depth and density.
- Calculate the pressure at a point in a liquid using $p = h\rho g$.
- Calculate the resultant forces produced by pressure differences.
- Investigate the factors that affect whether an object will float or sink.

Student reflection

Encourage students to reflect on whether there were any checkpoint questions they found difficult or straightforward as well as their preparation for the checkpoint assessment, including how they revised, what they revised, the time spent on revision, and what areas they could improve upon. For more strategies, see the Metacognition spread.

Intervention activity: Use the equation for pressure to determine pressure, area, or force.

The defining equation for pressure, $p = \frac{F}{A}$, requires significant cognitive effort from students, bringing together two concepts – force and area – to create a new idea about pressure (which often comes with misconceptions). It is useful, therefore, to spend time defining force and area separately, before bringing them together to explore how forces behave when applied over large and small areas. Use language such as 'applied force' and 'area over which the force is applied', identifying (e.g., shading or highlighting) surface areas as appropriate. Pens, pencils, compasses, and other pointed classroom objects are other examples students could find to illustrate forces applied over very small areas.

Where possible, give students tactile experience of large and small pressures generated by applying forces, for instance, comparing the penetration depth into soft sand for a cylinder (pushing its flat surface against the sand) and a cone. Students should recognise that for the same applied force, the cone is much easier to push into the sand. Explain this effect using this framing – from the 'sand's perspective', a larger pressure has been exerted, so it has no choice but to 'give way' to the piercing cone. Students should explain the sand's experience of a larger pressure by identifying, with reference to the equation, which quantity has not changed (the applied force) and which quantity has changed (the area over which the force is applied, which is much smaller for the cone). Following this, they should be confident applying the equation with simple values for force and area.

Ensure they recognise the pascal as the unit for pressure, which is equivalent to the Newton per m^2.

Intervention activity: Explain why liquid pressure increases with depth and density.

Revisit the demonstration of liquid pressure at differing depths using a pressure column and water, or a plastic bottle as in the pressure tests experiment in P11.2. This demonstration is a quick and effective illustration of increasing liquid pressure at larger depths. To reinforce this idea, ensure students can recognise the difference between high-pressure and low-pressure fluid ejection (e.g., use a short hose connected to a tap, or the tap itself). They should quickly conclude that liquid pressure increases with depth (which confusingly, is sometimes called 'height' from the surface of the liquid). On a sketch of the pressure column, or plastic bottle, ensure students can label this depth/height from the surface of the liquid.

Once students can state the effect of increasing depth on liquid pressure, they should be given the fact that the density of the fluid also increases the liquid pressure. Alongside a visual explanation, link this to the weight of fluid above different horizontal lines to the amount of pressure experienced along that surface. A sketch of underwater marine life, spread across different depths, is a useful visual aid to create a narrative about the liquid pressure experienced by different creatures, submarines, and so on. If additional support is needed, create sketches of the cross-section of fluid above each object. A marine diver, for instance, can be shown to have far less volume of water (i.e., less weight) acting above them, than a deep-sea whale.

Intervention activity: Calculate the resultant forces produced by pressure differences.

Since liquid (water) pressure is larger at greater depths, there is a pressure difference on surfaces of submerged objects. This is like a tug-of-war at both ends of an object! Use this to frame students' ideas about ensuing resultant forces; a large sketch of a cylinder submerged in water, such as that shown in P11.4 Figure 2 in the Student Book, is a great starting point to visualise this pressure difference. Tweak this image by exaggerating the size difference between the arrows representing pressure (strictly, pressure is not a vector quantity but emphasise this is a visual to help imagine the size difference).

Resultant forces

Use simple force values acting at the upper and lower surface of the submerged cylinder, for example, 100 N and 200 N, and model examples of calculating the resultant between them. Link these examples explicitly to students' prior knowledge about unbalanced forces, resultant forces, and the effect of resultant forces (i.e., acceleration). Students should be able to determine the direction of the resultant force on the cylinder in a variety of circumstances, coming to the realisation that the resultant force is *always* upwards. Ensure they can explain this in terms of the fact that the force (per unit of area) exerted at the cylinder's lower surface is *always* larger than that exerted on its upper surface. Identify this upward resultant force as the upthrust. Finally, ensure there is no misconception that the object, therefore, *always rises* to the surface: the resultant force in this context is due to pressure differences in the liquid. There is *another* force acting on the submerged object – its weight – which has nothing to do with the liquid it is submerged in. Hence, to determine whether it sinks or floats, students should compare the *total* upward force (the upthrust) to the *total* downward force (the weight).

4 Waves, electromagnetism, and space

Introduction to unit

This unit will enable students to develop their knowledge of waves and electromagnetism. A more quantitative model of waves will be studied, with students using calculations relating to wave speed, frequency, and wavelength. In terms of electromagnetism, students will determine the forces acting on a current-carrying wire and calculate its magnitude, before applying this to uses such as motors and transformers. In terms of space, students will go beyond the Solar System to look at the nature of the Universe, its formation, its composition, and its expansion (physics only). The evidence for this model of the Universe will be covered in some depth.

Assessment

Suggested minimum guided learning hours: 32 hours

Exam: Paper 2

Prerequisite knowledge from KS3 (chapter references from *Oxford Smart* Activate)

P3.3.1 Reactivate: Waves, sound, and light

- Waves on water act as undulations which travel through water with transverse motion; these waves can be reflected and add or cancel – superposition.

P3.3.3 Apply: Sound and its applications

- Frequencies of sound waves, measured in hertz (Hz); echoes, reflection and absorption of sound.
- Sound needs a medium to travel; the speed of sound is different in air, water, and solids.

P3.3.5 Apply: Light and colour

- The similarities and differences between light waves and waves in matter.
- The transmission of light through materials: absorption, diffuse scattering, and specular reflection at a surface.

P3.3.5 Apply: Applications of reflection and refraction

- Use of ray model to explain imaging in mirrors, the pinhole camera, the refraction of light and action of convex lens in focusing (qualitative).
- Colours are formed by the different frequencies of light, and white light can be refracted by prisms (qualitative only).

P3.4.6 Reach: Electromagnetism and induction

- The magnetic effect of a current, electromagnets, D.C. motors (principles only).
- Gravity force, *weight* = *mass* × *gravitational field strength (g)*, on Earth g = *10 N/kg*, different on other planets and stars; gravity forces between Earth and Moon, and between Earth and Sun (qualitative only).

AQA GCSE Physics specification links

4.6 Waves

- 4.6.1 Waves in air, fluids, and solids
- 4.6.2 Electromagnetic waves
- 4.6.3 Blackbody radiation (physics only)

4.7 Magnetism and electromagnetism

- 4.7.1 Permanent and induced magnetism, magnetic forces, and fields
- 4.7.2 The motor effect
- (H) 4.7.3 Induced potential, transformers, and the National Grid (physics only)

4.8 Space physics (physics only)

- 4.8.1 Solar system; stability of orbital motions; satellites (physics only)
- 4.8.2 Red-shift (physics only)

Identity, awe, and wonder

Students need to know that all waves transfer energy without any matter being transferred. Electromagnetic waves reach us from the Sun via the vacuum of space, with wavelengths shorter than ultraviolet light being absorbed by molecules in the Earth's atmosphere. Radio waves can easily pass through walls to bring us TV and radio signals, but they cannot penetrate the Earth's atmosphere to be received by satellites. Infrared and visible are not transmitted by walls – they are absorbed or reflected, but they can be used to send information down optical fibre cables. Microwaves can be used for satellite communications, and they are absorbed by water molecules to cook food. Humans can only see using a tiny band of the electromagnetic spectrum called visible light, whereas bees and birds can see ultraviolet radiation. Migratory birds make use of electromagnetic fields whereas dolphins and bats use ultrasound to locate objects they cannot see.

Required practicals

AQA required practicals	Practical skills	Topic
Required practical 8: Waves Make observations to identify the suitability of apparatus to measure the frequency, wavelength, and speed of waves in a ripple tank and waves in a solid and take appropriate measurements.	AT 4 – make observations of waves in fluids and solids to identify the suitability of apparatus to measure speed, frequency, and wavelength.	P12.4
Required practical 9: Light Investigate the reflection of light by different types of surface and the refraction of light by different substances.	AT 4 – make observations of the effects of the interaction of electromagnetic waves (light) with matter. AT 8 – make observations of waves in fluids and solids to identify the suitability of apparatus to measure the effects of the interaction of waves with matter.	P14.3
Required practical 10: Radiation and absorption Investigate how the amount of infrared radiation absorbed or radiated by a surface depends on the nature of that surface.	AT 1 – use appropriate apparatus to make and record temperature accurately. AT 4 – make observations of the effects of the interaction of electromagnetic waves with matter.	P13.2

Working scientifically

Students will give examples as to how scientific models and theories have changed over time, drawing on a range of data collected from experiment and observation. They will describe and explain a variety of technological applications of science related to the material in the unit.

Students will plan, collect, and analyse data related to waves and radiation, evaluating their methods and commenting on any predictions and conclusions.

They will present their data and observations using appropriate methods and comment on any patterns or trends they have noticed either numerically or in graphical form.

Maths skills

Students will be expected to recognise and use fractions, decimals, and percentages in calculations, and represent very large and small numbers in standard form. They will rearrange formulae and substitute values in order to calculate the size of physical quantities, using appropriate units.

Students will use angular measures in degrees. Students will also be expected to calculate areas and volumes.

Literacy links

The practical investigations that students will encounter in this unit will serve as an opportunity for writing skills, such as the writing of a step-by-step method. Other literacy-based activities in this unit can be found in P13.5, P14.5, P15.3, P16.3, and P16.5.

P12 Wave properties

P12.1 The nature of waves

AQA GCSE Physics specification points

4.6.1.1 Transverse and longitudinal waves

4.6.2.1 Types of electromagnetic waves

***Smart Curriculum* Narrative**

Waves

Theme/subtheme

Waves/Types of waves

Disciplinary knowledge

- WS 1.2, 2.2

Practical

Wave circus

	Learning outcomes		
Learning objectives	**Developing Aiming for 3**	**Secure Aiming for 5**	**Extending Aiming for 8**
Describe evidence that for both sound and ripples on water it is the wave that travels and not the air or water	State that waves can transfer energy and information without the transfer of matter	**Describe evidence that for both sound and ripples on water it is the wave that travels and not the air or water**	Sketch diagrams to show the movement of a single particle about its fixed position during a transverse wave and its movement during a longitudinal wave
Compare transverse and longitudinal waves in terms of direction of vibration and propagation	Identify waves as either transverse or longitudinal	**Compare transverse and longitudinal waves in terms of direction of vibration and propagation**	Explain the features of a longitudinal wave in terms of compressions and rarefactions by using a particle model
Compare electromagnetic and mechanical waves	Identify waves as either mechanical or electromagnetic	**Compare electromagnetic and mechanical waves**	Compare mechanical waves and their particulate nature with electromagnetic waves and their field oscillations

Tier 2 vocabulary	Tier 3 vocabulary
energy, gravitational, matter, motion, parallel, particle, perpendicular	displaced, electromagnetic, longitudinal, mechanical, *medium*, oscillate, seismic, transverse, wave

Lesson plan

Common misconception

Students may think that particles in a wave travel with the wave. Tell students that when a (mechanical) wave passes through a medium, it causes the particles that make up that medium to oscillate about their original positions because waves only transfer energy, not matter.

Reactivate knowledge

1 Give three examples of waves.

2 Give an example of a wave that can travel through space.

3 How do you make a wave on a rope?

Answers: 1 Any electromagnetic, water, sound *2* Any electromagnetic wave *3* By moving one end in different directions

Trigger interest

Wave theme (5 mins)

Show students a montage of different waves and ask them what the common theme is. To extend the challenge, ask them to identify as many of the waves as they can. Include a range of examples they will need to know, for example, sound, water, seismic, electromagnetic, waves on a spring or rope, as well as a simple demonstration illustrating transfer of information, for example, a remote control transferring information via infrared waves. Inspire some additional interest by using some potentially unfamiliar examples, for example, gravitational waves transferring energy across the universe.

Exposition

Mechanical and non-mechanical waves (20 mins)

This lesson is reliant on learning and employing a variety of technical terms, so a display of keywords to refer to throughout the lesson will be useful, for example, transverse, longitudinal, mechanical, electromagnetic, perpendicular, parallel, and oscillate.

Show students a clip of a small object on the surface of a pond or body of water as ripples pass. They should note that the object is not permanently displaced – it only oscillates about its original position as waves pass underneath. Use this to define a wave, before categorising examples as transverse or longitudinal waves. The PhET simulations for waves on a string, and sound waves through air, are useful interactive tools that help to illustrate the differences between transverse and longitudinal waves, allowing students to focus on individual particle oscillations vs. wave motion (energy transfer). Ensure that students can recognise and contrast the perpendicular and parallel directions of motion.

Use a light source to demonstrate electromagnetic waves. If available, a short demonstration of objects illuminated by UV light works well. Link to students' prior knowledge about energy to reinforce that these waves act as an energy pathway (i.e., they heat surfaces) so, they too, transfer energy – *not* matter or material. Use the Sun to emphasise that electromagnetic waves need no medium to travel through – no particles are oscillating as the waves travel, hence the difference between mechanical and electromagnetic waves.

Main activity

Wave circus (25 mins)

Create a set of wave demonstrations and instructions for students. Ensure a variety of examples are displayed, for example, slinky springs, oscilloscopes (for sound waves), ripple tanks, and a laptop or tablet to access an appropriate simulation of transverse and longitudinal waves.

Review and practice

The same but different (5 mins)

Interactive activity where students match key words from the lesson to their definitions, complete a description of longitudinal and transverse waves, then identify true and false statements about longitudinal and transverse waves.

P12.2 The properties of waves

AQA GCSE Physics specification points

4.6.1.2 Properties of waves

Smart Curriculum Narrative

Waves

Theme/subtheme

Waves/Wave properties

Disciplinary knowledge

- MS 1c, 3b, 3c

Practical

Measuring the speed of ripples

Learning objectives	Learning outcomes: Developing Aiming for 3	Learning outcomes: Secure Aiming for 5	Learning outcomes: Extending Aiming for 8
Describe the motion of a wave using the terms frequency, wavelength, amplitude, and period	Identify wavelength and amplitude on a diagram	**Describe the motion of a wave using the terms frequency, wavelength, amplitude, and period**	Calculate the period of a wave from its frequency
Calculate the speed, frequency, or wavelength of a wave using the wave equation	Calculate the speed of a wave using the wave equation	**Calculate the speed, frequency, or wavelength of a wave using the wave equation**	Explain how the wave speed equation can be derived from fundamental principles
Describe how to measure the speed of a wave in water	Measure the wavelength and frequency of waves in a ripple tank to determine speed of waves	**Describe how to measure the speed of a wave in water**	Describe the sources of uncertainty in measuring the speed of a wave

Tier 2 vocabulary	Tier 3 vocabulary
distance, energy, parallel, particle, speed, time	amplitude, displacement, frequency, *medium*, oscillate, peak, *period*, transverse, *trough*, wavelength

Lesson plan

Reactivate knowledge

1 What is the amplitude of a wave?

2 Is a wave on water an example of transverse or longitudinal wave?

3 How do you calculate speed?

Common misconception

Students may confuse amplitude with height. The amplitude of a wave is the maximum displacement of a wave from its undisturbed – or equilibrium – position. This is half of the vertical distance between the peak and the trough.

Answers: 1 Distance from the middle to the top of the wave *2* Transverse *3* Speed = distance ÷ time

Trigger interest

Up to speed (10 mins)

Use the interactive to perform some simple speed calculations to remind students of this work from Key Stage 3. Ensure that the correct units are used for speed, distance, and time.

Extend this exercise by requiring rearrangement of the equation and using more challenging data, for example, distance and time in non-SI units.

Exposition

Sensing and describing waves (15 mins)

Split students into three groups corresponding to three senses (touch, sight, hear). Each group should suggest at least two waves that are sensed in their category, e.g., visible light is seen, infrared waves are felt (as 'heat') or seen (by animals, like snakes). Taste and smell, potentially more challenging, could be provided as an extension challenge to interested students. The purpose of this activity is to convince students that waves are all around us, and we interact with them all the time.

Use a rope or slinky spring to demonstrate the basic shape of a transverse wave. The online simulation from PhET – Waves on a string allows for a more detailed look at the key characteristics of a wave. Use this to define and discuss the amplitude and wavelength, pausing and annotating where appropriate.

Then, continue the analysis by changing the frequency and showing the number of waves passing a fixed point per second on the rope, slinky spring, or simulation. Introduce the mathematical relationship between frequency and period and give students an opportunity to perform a few simple calculations using mini-whiteboards

Main activity

Measuring the speed of ripples (20 mins)

If available, show the basic operation of a ripple tank (a waterproof tray can be used as an alternative, alongside a simulation or video clip), focusing observations on the motion of a single wave crest. Refer to a sketch of the parallel wave crests to ensure students can identify the wavelength (this is difficult to measure using a ruler directly, so wherever possible, taking a photograph with a ruler attached to the tank/tray makes this easier).

Students should appreciate that crests move across the tank or tray with a particular speed, and that changing the frequency (e.g., of the bobbing ruler) affects the wavelength, while the speed stays the same.

Then, introduce the wave speed equation, modelling a calculation, after which, students can use the ripple tanks/waterproof trays to investigate the wave equation independently.

Review and practice

Kinaesthetic maths challenge (10 mins)

Provide groups of five students with cards labelled with '×', and '=', and a set of values of frequency and wavelength. Students must identify frequencies and wavelengths and arrange themselves into 'living equations' by standing in order in their group to make a correct equation. Their classmates can verify their equations by performing the calculations.

Provide additional challenge by giving chosen groups more difficult values, for example, using prefixes like kHz and MHz, and ask them to explain their reasoning to their classmates where appropriate.

P12.3 Reflection and refraction

AQA GCSE Physics specification points

(H) 4.6.2.2 Properties of electromagnetic waves 1

Smart Curriculum Narrative

Waves

Theme/subtheme

Waves/Wave properties

Disciplinary knowledge

- WS 1.2
- MS 5a, 5b

Practical

Reflection and refraction tests

Learning objectives	Learning outcomes: Developing Aiming for 3	Learning outcomes: Secure Aiming for 5	Learning outcomes: Extending Aiming for 8
Describe reflection and refraction of waves	Observe reflection and refraction of a ripple tank wave	**Describe reflection and refraction of waves**	Compare refraction and reflection
Describe refraction as being caused by the change in velocity of the waves in different substances	Draw ray diagrams to show refraction of a wave as it moves from one medium into another	**Describe refraction as being caused by the change in velocity of the waves in different substances**	Use wave front diagrams to explain refraction at a boundary between two different substances
Describe how whether a wave is transmitted, reflected, absorbed, or refracted depends on both the substance and the wavelength of the wave	State that a wave can be transmitted, reflected, absorbed, or refracted when it is directed at a substance	**Describe how whether a wave is transmitted, reflected, absorbed, or refracted depends on both the substance and the wavelength of the wave**	Explain partial absorption as a decrease in the amplitude of a wave and therefore the energy carried

Tier 2 vocabulary	Tier 3 vocabulary
angle of incidence, parallel, perpendicular, speed	absorb, air-water boundary, boundary, *medium*, ray, reflection, refraction, transmit

Lesson plan

Reactivate knowledge

1 What is the law of reflection?

2 What is refraction?

3 Name two things that can happen when a wave crosses a boundary between two different mediums.

Common misconception

Students may think refraction of waves refers to the change of direction of waves as it enters a different medium. It is actually waves changing speed in a new medium. This sometimes results in a change of direction, but not always – if a wave crosses a boundary to a new medium perpendicular to the boundary, it does not change direction, but it still refracts (changes speed).

Answers: 1 The angle of incidence = the angle of reflection *2* The change in direction of a wave as it enters or leaves a medium *3* Reflection, transmission, absorption

Trigger interest

Money appears out of nowhere (10 mins)

Use a mug and a coin to demonstrate the effects of refraction with small groups in turn. Place the coin at the bottom of the mug and ask onlooking students to move backwards until the moment the coin disappears out of view. Slowly pour water into the mug, and the coin will reappear.

Ask students to discuss and explain their observations, in terms of refracting light waves.

This is a useful way to survey prior knowledge about these effects.

Exposition

Demonstrating reflection and refraction (15 mins)

Show students a large diagram of a reflected light ray, such as Figure 1 in the Student Book. Emphasise that light is made up of waves, and distinguish between the parallel wavefronts (which are the crests of each wave) and the ray (which indicates the direction of travel). This is not always clear to students. Use this diagram to define reflection.

Next, demonstrate the operation of a ripple tank, if available (alternatively, a plastic tray with water is effective), focusing on reflection at its flat edge. The students can introduce a barrier during the practical to make these observations clearer, as well as exploring reflection at different angles of incidence.

Finally, show students the effect of refraction by changing the depth of the water. It is important to emphasise the change in speed of the wavefront, which *can* result in a change in direction. Explain this effect using wavefronts and wavelets; a simple diagram (e.g., Figure 3 in the Student Book) is a useful reference. As an additional challenge, students could observe and describe the fact that wavelength changes but frequency stays constant.

Main activity

Reflection and refraction tests (20 mins)

Students use the ripple tank or tray with water to explore the effects of reflection and refraction. It might be useful to supplement these observations with simulations, for example, PhET – Bending Light, which allows students to manipulate rays of light for both reflection and refraction.

Circulate amongst the groups, and provide prompts to help students verbalise their observations. Ask, *"What are the factors that determine what happens to a wave when it reaches a surface?"*

Explore the various options with small groups (reflect, refract, transmit, absorb) explaining that the outcome is dependent on both the substance itself, and on the wavelength of the incoming wave. Highlight the heating effect on the substance in the case of absorption and partial absorption – reducing the amplitude of the incident wave and hence the energy transferred. Support students by providing a list of keywords and short definitions. Ray diagrams, wherever possible, are helpful visual aids.

Review and practice

Money appears out of nowhere, explained (10 mins)

Display a ray diagram of the coin in the mug, with and without water, and including an observing 'eye'. The refracted ray, going into the observing eye, should be clearly labelled, as should the air–water boundary.

Students should discuss and explain the effect by referring to the diagram and using the keywords from the lesson. Scaffold questioning as appropriate, ensuring they explain refraction in terms of light waves changing speed as they leave water and enter air, resulting in a change of direction toward the observing eye.

P12.4 More about waves

AQA GCSE Physics specification points

4.5.6.1.2 Speed

4.6.1.2 Properties of waves

Smart Curriculum Narrative

Waves

Theme/subtheme

Waves/Sound

Disciplinary knowledge

- MS 1c, 3b, 3c

Required Practical 8

Waves

	Learning outcomes		
Learning objectives	**Developing Aiming for 3**	**Secure Aiming for 5**	**Extending Aiming for 8**
Describe how to measure the frequency and wavelength of a wave on a string to determine the speed of the wave	Measure the frequency and wavelength of a wave on a stretched string	**Describe how to measure the frequency and wavelength of a wave on a string to determine the speed of the wave**	Evaluate the suitability of apparatus for measuring the frequency, wavelength, and speed of waves
Describe how sound waves are created and how they move through air	State that sound waves require a medium to travel in	**Describe how sound waves are created and how they move through air**	Apply understanding of a particle model and longitudinal waves to explain the difference in speed of sound in gases and solids
Describe how to accurately measure the speed of sound in air	State a method for measuring the speed of sound	**Describe how to accurately measure the speed of sound in air**	Evaluate data from speed of sound experiments to discuss the range of possible speeds for sound and sources of error

Tier 2 vocabulary	Tier 3 vocabulary
oscilloscope, speed, *stationary*	frequency, *medium*, oscillate, vacuum, vibration, wave, wavelength

Lesson plan

Reactivate knowledge

1 What is an echo?

2 How are speed, frequency and wavelength related?

3 What is the wavelength of a wave?

Common misconception

Students may think high frequency/pitched sounds travel faster than low frequency/pitched sounds. Respond that the properties of the medium is the only thing that affects the speed of sound waves through it. Frequency (or the pitch of the sound) does not alter the speed at which the wave travels.

Answers: 1 A reflection of a sound wave *2* speed = frequency × wavelength *3* The distance from a point on the wave to the equivalent point on the next wave

Trigger interest

Silent sound? (5 mins)

Demonstrate that sound waves can't travel through a vacuum. If available, use a vacuum pump connected to a bell jar to slowly remove the air and ask students to explain their observations. A video clip of this demonstration works well as an alternative.

Students should conclude that sound waves require a medium to travel through – or equivalently, that vibrations are needed to produce sound.

Exposition

Sound waves need a medium (15 mins)

If available, show a loudspeaker vibrating, or use a video clip. Use a simulation or animation to show how a loudspeaker causes air particles to vibrate – again illustrating the need for a medium.

Give students a selection of images of land and sea animals that utilise sound waves for navigation, hunting, or communication. Encourage ideas that elaborate on the images shown and come to a consensus that all sound waves require a medium to travel through. Students might be surprised that sound can travel through solid substances, for example, elephants use sound vibrations through solid earth to communicate, and snakes' inner ears are connected to their jawbone to enhance their hunting abilities.

Main activity

Waves (25 mins)

Students set up stationary wave patterns measuring both frequency (from the signal generator) and the wavelength (with the tape measure). They do not need to know the details of how stationary waves form, they only need to measure their wavelength. The method used to measure speed, by measuring frequency and wavelength, is the key learning from this required practical.

If there is time, students can investigate waves in a liquid using the practical guidance. This practical uses simple ripples which can be produced with a ruler or with a vibrating bar in a ripple tank.

Review and practice

Oscilloscope solutions (10 mins)

Students use the interactive to match a set of problems encountered when using the oscilloscope to appropriate solutions, for example, a trace where the waves' peaks are too close together – the solution is to reduce the time base.

Ⓗ P12.5 Sound waves

AQA GCSE Physics specification points

Ⓗ 4.6.1.4 Sound waves

Smart Curriculum Narrative

Waves

Theme/subtheme

Waves/Sound

Disciplinary knowledge

- MS 1c, 3b, 3c, 4a

Practical

Investigating different sounds

	Learning outcomes		
Learning objectives	**Developing Aiming for 3**	**Secure Aiming for 5**	**Extending Aiming for 8**
Outline the structure of the human ear in terms of transfer of waves and vibrations	State that microphones and the ear convert sound waves into vibrations	**Outline the structure of the human ear in terms of transfer of waves and vibrations**	Explain how the structure of the human ear helps transmission of sound
Explain why the human ear has a limited range of frequencies it can detect	State the range of frequencies that humans can hear	**Explain why the human ear has a limited range of frequencies it can detect**	Explain why although the speed of sound changes as it enters the different substances in the ear the frequency remains the same
Describe how high frequency sound waves can be used to measure depths of the sea bed	State that high frequency sound waves can be used to measure depths of the sea bed	**Describe how high frequency sound waves can be used to measure depths of the sea bed**	Calculate the depth of an object using data

Tier 2 vocabulary	Tier 3 vocabulary
distance, oscilloscope, particle, sine wave, speed, time	absorption, audible, auditory, boundary, echo, frequency, *medium*, *pitch*, receiver, transmitter

Lesson plan

Reactivate knowledge

1 How do the vibrations of high pitch sounds compare with low pitch ones?

2 How does a sound wave travel?

3 How does the speed of sound compare in solids and gases?

Common misconception

Students may think sound can't travel through solids and liquids. Sound waves are actually vibrations that travel through a medium (a substance - solid, liquid, or gas). Any vibrating object in contact with the medium makes its particles vibrate, which in turn causes the adjoining particles to vibrate, and so on. Travelling vibrations can cause human eardrums to vibrate, which our brains interpret as sound.

Answers: 1 Vibrations are faster in high pitch sounds *2* Particles in the medium vibrate parallel to the direction of the wave *3* Speed of sound is fastest in solids

Trigger interest

Frequency of human hearing (5 mins)

Use a loudspeaker and signal generator to demonstrate the range of audible frequencies for human hearing. Video clips of similar demonstrations work well if the equipment is not available. Ensure the volume is not too loud and avoid sudden noises.

Ask students to raise their hands to confirm they can hear a signal. Increase the frequency of the signal gradually, ensuring students can see the value of the frequency on the signal generator. Students should lower their hands as soon as they are unable to hear the sound. Continue to increase the frequency until it is beyond everyone's hearing range.

Where needed, adapt this demonstration to include students who may have additional auditory needs, for example, ensuring no sudden loud noises, or a very clear (audible) signal. It might be appropriate to incorporate a visual component to this demonstration using an oscilloscope, highlighting both audible and visible representations of the upper limits to hearing-frequency.

Exposition

Inside the ear (15 mins)

Provide students with a clear diagram of the human ear to annotate. They should identify the key features, focusing on the idea of absorption and reflection of sound waves at boundaries to explain why not all sounds can be detected.

Reinforce ideas covered in P12.4 about how sound waves propagate in solids compared to gases, ensuring students appreciate how closely packed particles will allow sound waves to travel more quickly.

Main activity

Investigating different sounds (25 mins)

Students explore the simple waveforms generated by a variety of objects, finding traces using oscilloscopes and microphones.

Circulate groups and have some creative fun with the demonstrations, for example, see which of the students can form the 'perfect note' by singing into a microphone and judging the resulting sine wave on the oscilloscope. Ensure that ideas about loudness and amplitude, and pitch and frequency, are discussed, encouraging students to tinker with the time-base setting of the oscilloscope.

As before, ensure this activity is adapted for students with additional auditory needs. Since this activity relies on the perception of sound, care must be taken to provide visual instruction about the vibrations for students with a hearing impairment. Wherever possible, encourage students to think about the 'see-and-feel' aspect of the equipment (i.e., tuning forks, loudspeakers) which will allow students to benefit more fully from the practical.

Review and practice

Echo sounding (10 mins)

Show a simple ray diagram of a ship using echo sounding off the seabed, labelling the transmitter and the receiver. On mini-whiteboards, students should write down the equation that would determine the distance to the seabed.

Scaffold this task as appropriate, reminding students of the equation linking (wave) speed, distance, and time.

Extend this activity by asking students to calculate the distance to the seabed given values for the speed of sound in water, and the time taken for a transmitted signal to be received.

H P12.6 The uses of ultrasound

AQA GCSE Physics specification points

H 4.6.1.5 Waves for detection and exploration

Smart Curriculum Narrative

Waves

Theme/subtheme

Waves/Sound

Disciplinary knowledge

- WS 1.4
- MS 3b, 3c, 4a

Practical

Investigating ultrasound

	Learning outcomes		
Learning objectives	**Developing Aiming for 3**	**Secure Aiming for 5**	**Extending Aiming for 8**
Explain what is meant by ultrasound	State that sound can have any frequency	**Explain what is meant by ultrasound**	Explain why pulses of ultrasound are used rather than a continuous signal when using ultrasound to measure distances
Describe how the partial reflection of ultrasound at a boundary between two substances is used to form an image	Describe what happens to a wave as it crosses a boundary between two different substances	**Describe how the partial reflection of ultrasound at a boundary between two substances is used to form an image**	Explain the advantage of ultrasound over X-rays in medical imaging
Describe how reflections of an ultrasound pulse can be used to detect a flaw in a metal bar	Identify the reflections of an ultrasound pulse on a diagram of an oscilloscope trace	**Describe how reflections of an ultrasound pulse can be used to detect a flaw in a metal bar**	Calculate the positions of objects or flaws in metal objects using data from an oscilloscope trace

Tier 2 vocabulary	Tier 3 vocabulary
oscilloscope, particle	frequency, *medium*, pulse, reflect, ultrasound

Lesson plan

Reactivate knowledge

1 What is the range of human hearing?

2 Sometimes when you look through a window, you can see through it and also see your reflection. Why is that?

3 How can a ship use high frequency sound to measure the depth of the water?

Common misconception

Students may think sound can't travel through solids and liquids. Sound waves are actually vibrations that travel through a medium (a substance - solid, liquid, or gas). This medium might be a solid, liquid or gas. Any vibrating object in contact with the medium makes the its particles vibrate, which in turn causes the adjoining particles to vibrate, and so on. Travelling vibrations can cause human eardrums to vibrate, which our brains interpret as sound.

Answers: 1 20 to 20 000 Hz *2* Some light transmitted and some reflected *3* Send a pulse of sound, and time how long before the pulse returns. Use $d = \frac{1}{2}vt$ to calculate depth.

Trigger interest

Mystery scans (5 mins)

Show students a collection of ultrasound scans of foetuses and some of the organs in the human body. Ask them to identify what they see. Images can be differentiated appropriately.

Exposition

Echo (15 mins)

Revisit the demonstration used in P12.5. With a frequency well above the range of human hearing displayed on the oscilloscope, ask students to suggest a definition for ultrasound. Use bats as an example of a species that uses ultrasound it to locate prey. Walk students through a simple calculation (e.g., the distance to a cliff) to demonstrate how these waves can be used to determine the distance to objects.

Ensure students can recall the speed of sound in air as 330 m/s.

Extend this calculation by exploring how ultrasound waves are absorbed and reflected by human tissue and bone to create images, for example, for ultrasound travelling in fat (1450 m/s) or bone (4000 m/s). Before starting the practical, use the interactive to get students practising how to calculate the time it would take for an ultrasound wave to be reflected.

Main activity

Investigating ultrasound (25 mins)

Students investigate ultrasound directly using an oscilloscope and a signal generator. If time permits, additional materials can be used to determine whether ultrasound waves can be reflected and absorbed by different materials.

As before, ensure that the activity is accessible for students who may have additional auditory needs, such as ensuring no sudden loud noises, and a very clear (audible) signal. It might be effective to encourage collaboration between pairs of students such that they communicate how they perceive sound waves at high frequencies, for example, by sight, touch, and/or hearing.

Review and practice

The wonders of ultrasound (10 mins)

Provide students with three case studies about how ultrasound is used in industry. E.g., the use of ultrasonic cleaning in dentistry, pest deterrents in agriculture, and in marine biology to study dolphins. Students can select one of the case studies and summarise the information in two or three bullet points, or answer some questions.

Ⓗ P12.7 Seismic waves

AQA GCSE Physics specification points

Ⓗ 4.6.1.5 Waves for detection and exploration

Smart Curriculum Narrative

Waves

Theme/subtheme

Waves/Seismic waves

Disciplinary knowledge

- WS 1.1, 1.4
- MS 1b

Practical

The seismometer

	Learning outcomes		
Learning objectives	**Developing Aiming for 3**	**Secure Aiming for 5**	**Extending Aiming for 8**
Describe the internal structure of the Earth	State that the Earth 's structure consists of the inner core, outer core, mantle, and crust	**Describe the internal structure of the Earth**	Explain the difficulties of exploring inside the Earth
Compare the properties of P and S waves	Describe the properties of P and S waves	**Compare the properties of P and S waves**	Calculate the speed of different types of seismic waves or distance to epicentre using data provided
Describe how P and S waves provide evidence for the structure of the Earth	State that P and S waves provide evidence for the structure of the Earth	**Describe how P and S waves provide evidence for the structure of the Earth**	Explain in detail how to use the detection of seismic waves to determine the structure of the Earth

Tier 2 vocabulary	Tier 3 vocabulary
speed	seismic, seismometer, shadow zone, ultrasound, wave

Lesson plan

Reactivate knowledge

1 What is the difference between longitudinal and transverse waves?

2 Sound waves are longitudinal waves. Which state of matter do they travel fastest in?

3 What happens to waves when they enter another medium and change speed?

Answers: 1 Longitudinal oscillations/vibrations are parallel to direction of energy transfer and in transverse waves they are perpendicular to direction of energy transfer *2* Solid *3* Waves are refracted

Common misconception

Students may think seismic waves travel across the surface of the Earth causing direct damage to buildings. Respond that seismic waves are generated by sudden movement of the Earth's crust, and travel through the core. They are able to shake the surface of the Earth, and the buildings and structures built on it.

Trigger interest

Earthquakes (10 mins)

Set up an activity to explore students' prior knowledge about earthquakes. Some students may refer to recent noteworthy earthquakes around the world, or the safety features that engineers must consider for buildings in earthquake-prone regions. Create three stations in the classroom, for example, 'real-world examples', 'causes' and 'effects', for student pairs to move around and discuss. Ask targeted questions as appropriate.

Support this activity by providing visual prompts at each station. For example, newspaper articles or short fact sheets to aid conversation.

Exposition

Seismic waves (15 mins)

Outline the structure of the Earth using a large, labelled diagram. Students should be able to identify the different layers, and appreciate that earthquakes move layers of rock, releasing shock waves (seismic waves) that provide information about the Earth's internal structure. Students should contrast this principle to the use of ultrasound waves, which provide information about the internal structures of living organisms.

Discuss the different types of seismic waves and the media they can and cannot travel through. Link the speed of the wave to the time of arrival at the detector. This evidence can be difficult to understand, so ensure the focus is on the meaning of the 'shadow zone', which demonstrates the existence of a liquid outer core.

Main activity

The seismometer (25 mins)

Students investigate the operation of a seismometer by building a simple model. This experiment involves a lot of construction, so ensure roles are split evenly in pairs or amongst groups.

As students complete the practical, circulate groups and discuss simple examples of using seismometer data to calculate the speed of different seismic waves, or examples of how data is used to determine the epicentre of earthquakes.

Review and practice

Make waves (5 mins)

Interactive activity where students choose the missing words to complete a summary on seismic waves and the evidence they provide for the structure of the Earth.

P12 Wave properties

Introduction to checkpoint intervention

The checkpoint intervention provides suggestions for a lesson to follow up the *P12 Wave properties* checkpoint assessment. Depending on the outcomes of the assessment, these suggestions can be used to move a class towards secure, to target specific learning outcomes, or to consolidate knowledge for students achieving secure outcomes. Use students' outcomes from the checkpoint assessment to address general misconceptions from the content covered. Be prepared to re-cover content as required; three key concepts from the chapter are covered in more detail below.

Checkpoint secure learning outcomes

- Compare electromagnetic and mechanical waves.
- Calculate the speed, frequency, or wavelength of a wave using the wave equation.
- Describe how whether a wave is transmitted, reflected, absorbed or refracted depends on both the substance and the wavelength of the wave.
- Describe refraction as being caused by the change in velocity of the waves in the different substances.
- Describe why sound waves travel more quickly in solids than they do in gases.
- Compare the properties of P and S waves.

Student reflection

Encourage students to reflect on whether there were any checkpoint questions they found difficult or straightforward as well as their preparation for the checkpoint assessment, including how they revised, what they revised, the time spent on revision, and what areas they could improve upon. For more strategies, see the Metacognition spread.

Intervention activity: Describe how whether a wave is transmitted, reflected, absorbed or refracted depends on both the substance and the wavelength of the wave.

A secure understanding of frequency and wavelength should precede activities that measure these quantities. Students will likely be able to describe how to measure these wave properties if they know what they're looking for. For wavelength, a selection of annotated images is effective.

Students should find patterns (e.g., the odd one out) of a collection of wavelength measurements, leading them to the definition and understanding of wavelength.

For frequency, use a variety of animations and simulations to demonstrate oscillations at low and high frequencies. A signal generator and oscilloscope combination are very effective; giving students an opportunity to explore the audible and visible effects of varying frequency, leading them to the definition of frequency in terms of oscillations or vibrations per second.

Frame the approach to measurement using the types of equipment that are needed to measure length (e.g., a metre ruler) and oscillations (i.e., something that can keep count every second).

Set up the waves on a string apparatus, asking students to measure the length of string (between the vibrator and the pulley). Adjust the frequency of the signal generator until there is a single loop on the string.

Draw a large sketch of the half-wavelength pattern and ask students to describe how they could use what they know to figure out the length of a full wavelength 'shape'. This is a subtle prompt to encourage students to 'complete' the wavelength by doubling the length of the string. The same system can produce a complete wave (instead of half) so they can be shown a 'whole wave' too.

The frequency of the oscillator can be read directly from the signal generator – model this for students, ensuring they verbalise this process. Finish this demonstration showing how the measurements are substituted directly into the equation to determine the speed of the waves on the string.

Intervention activity: Calculate the speed, frequency, or wavelength of a wave using the wave equation.

There are a variety of equation frameworks to help students use and rearrange equations. Equation triangles tend to be useful interventions to build student confidence at using equations. Model how this is used for the wave speed equation, starting with calculations of speed with simple whole numbers (without prefixes).

Explore what happens using the equation triangle if frequency and wavelength are swapped, that is: $v = f\lambda$ vs. $v = \lambda f$. Students should recognise that the order of multiplication doesn't matter.

Once students are confident calculating wave speed, model how to use the equation triangle to determine the frequency or wavelength. This is a rearrangement of the wave speed equation (this makes frequency, or wavelength, the subject of the equation).

Ensure students can see from the equation triangle that speed is always on the top of the fraction, that is, $f = \frac{v}{\lambda}$ or $\lambda = \frac{v}{f}$

Once students can use the equation frame to determine frequency and wavelength, introduce additional examples using prefixes, for example, km for wavelength, and GHz for frequency.

Intervention activity: Describe why sound waves travel more quickly in solids than they do in gases.

The fact that sound waves travel at a finite speed in air may not be obvious or properly understood by students. Demonstrate that sound requires a medium to travel through; this is the foundational concept. If available, a bell jar and vacuum pump is a great demonstration to illustrate this point (a video clip works well too). Students may be familiar with modern sci-fi films that depict the silence of epic battles in outer space. Use these demonstrations to explore the idea that we perceive sound by virtue of series of interlinked vibrations.

On Earth, these vibrations typically travel as waves through particles in air – again, a simulation or animation of this effect is useful. Ensure the source and detector (i.e., our ear drums) are specified. Provide a short gap-fill activity for students to summarise this concept, providing keywords that they should select as appropriate.

Once the conceptual basis of how sound waves travel is understood, pose students the question: *How fast do they travel?*

Gather estimations, scaffolding further questioning related to whether the medium of travel would affect the speed. Provide a list of sound speeds in different media and ask students to spot any patterns – they may need prompting to group the data into solids and gases. This should lead them to conclude that sound waves travel faster in solids than in gases.

Remind students of the particle structure of solids and gases (two contrasting images will help), link back to the idea of vibrations being passed along (particle-by-particle might be a useful simplification) to conclude why sound can travel through solids more quickly.

P13 Electromagnetic waves

P13.1 The electromagnetic spectrum

AQA GCSE Physics specification points

4.6.1.2 Properties of waves

4.6.2.1 Types of electromagnetic waves

Ⓗ 4.6.2.2 Properties of electromagnetic waves 1

Theme/subtheme

Waves/Electromagnetic waves

Practical

Introducing electromagnetic waves

***Smart Curriculum* Narrative**

Waves

	Learning outcomes		
Learning objectives	**Developing Aiming for 3**	**Secure Aiming for 5**	**Extending Aiming for 8**
Describe how electromagnetic (EM) waves transfer energy from a source to an absorber with examples	State that electromagnetic (EM) waves transfer energy from a source to an absorber	**Describe how electromagnetic (EM) waves transfer energy from a source to an absorber with examples**	Describe the relationship between the energy being transferred by an electromagnetic wave and the frequency of the wave
Identify the position of EM waves in the spectrum in order of wavelength and frequency	Name the groups of electromagnetic waves	**Identify the position of EM waves in the spectrum in order of wavelength and frequency**	Explain the patterns in wavelength and frequency across the electromagnetic spectrum
Calculate the frequency and the wavelength of an electromagnetic wave	State that all EM waves travel at the same speed in a vacuum	**Calculate the frequency and the wavelength of an electromagnetic wave**	Use standard form in calculations of wavelength, frequency, and wave speed

Tier 2 vocabulary	Tier 3 vocabulary
constant, electric field, energy, light, magnetic field, wave, radiation, radio, sound, speed, standard form	absorb, detect, electromagnetic spectrum, frequency, oscillating, reflect, source, transmit, transverse, *vacuum*, wavelength

Lesson plan

Reactivate knowledge

1 Is light a transverse or a longitudinal wave?

2 What are the colours of visible light in order of increasing wavelength?

3 How do you calculate the wave speed using frequency and wavelength?

Common misconception

Students may think radio waves travel at the speed of sound. Tell students that radio waves are not sound waves. They are part of the electromagnetic spectrum, which is a set of waves that all travel at the same speed through a vacuum (3×10^8 m/s). This is the speed of light. All waves on the electromagnetic spectrum, in this sense, only differ by their frequency and wavelength, so can all be considered as light waves.

Answers: 1 Transverse wave *2* Red, orange, yellow, green, blue, indigo, violet *3* Speed = frequency × wavelength

Trigger interest

Electromagnetic mural (5 mins)

Show students a collection of images containing examples of the electromagnetic spectrum. They should determine the common theme that unites the images.

There are various ways in which this activity can be extended in difficulty. There may be an 'odd one out' (e.g., a sound wave), or some images may be more abstract (e.g., a diagram of an oscillating electric and magnetic field). Scaffold students' ideas by asking for more in-depth suggestions. For example, the common theme might be their speed, the fact that all electromagnetic waves can travel through a vacuum, or that they are all transverse waves. Use this to scope students' prior knowledge.

Exposition

Introducing electromagnetic waves (15 mins)

Recap the nature of electromagnetic radiation compared with mechanical waves. Focus on the wide range of wavelengths of the spectrum, linking this to the name and effects of each part. Emphasise how small the region of visible light is, compared to the entire spectrum. Ensure that the transfer of energy by electromagnetic waves is covered.

Show the link between frequency and wavelength, that is, longer wavelengths have lower frequencies (and vice versa) as demonstrated by the wave speed equation when speed is constant. Students should attempt a few calculations of wavelength and frequency. Support this, using a calculation framework, where appropriate. Some students may need additional scaffolding using their calculators for very large and small numbers (i.e., when inputting and interpreting answers in standard form). Ensure all students have opportunities to use and interpret standard form here, for example, using the speed of light as 3×10^8 m/s.

Main activity

Introducing electromagnetic waves (25 mins)

In small groups, students can rotate through all of the experiments, exploring the properties of different electromagnetic waves. Alternatively, each group can carry out one experiment and then share and discuss their results.

As you circulate, ensure students can identify the source and absorber/detector of the electromagnetic wave under investigation.

For higher tier candidates, recap the fact that electromagnetic waves can be absorbed, reflected, and transmitted, leading students to conclude that all waves in the electromagnetic spectrum must behave in the same manner. Ask students to identify examples of this for different electromagnetic waves.

Review and practice

Electromagnetic calculations (10 mins)

Students complete the interactive where they can attempt further calculations of wavelength or frequency of electromagnetic waves.

P13.2 Light, infrared, microwaves, and radio waves

AQA GCSE Physics specification points

4.6.2.2 Properties of electromagnetic waves 1

4.6.2.4 Uses and applications of electromagnetic waves

Smart Curriculum Narrative

Waves

Theme/subtheme

Waves/Electromagnetic waves

Disciplinary knowledge

- WS 1.4
- MS 1b, 1c, 3b, 3c

Required Practical 10

Radiation and absorption

	Learning outcomes		
Learning objectives	**Developing Aiming for 3**	**Secure Aiming for 5**	**Extending Aiming for 8**
Describe how light, infrared, microwaves, and radio waves are used in a variety of scenarios	State some uses of light, infrared, microwaves, and radio waves	**Describe how light, infrared, microwaves, and radio waves are used in a variety of scenarios**	(H) Explain why a particular wave is suited to its application
Describe how the light, infrared, microwaves, and radio waves interact with matter	State any hazards of light, infrared, microwaves, and radio waves	**Describe how the light, infrared, microwaves, and radio waves interact with matter**	(H) Describe how the wavelength of light, infrared, microwaves, and radio waves affects how they interact with matter
Plan an investigation into the rate of cooling due to emission of infrared radiation	Measure the rate of cooling due to emission of infrared radiation	**Plan an investigation into the rate of cooling due to emission of infrared radiation**	Evaluate an investigation into the rate of cooling due to emission of infrared radiation

Tier 2 vocabulary	Tier 3 vocabulary
energy, *filter*, intensity, light, wave, radiation, radio, sound, speed	absorb, electromagnetic spectrum, emission, frequency, infrared, *microwave*, *vacuum*, wavelength

Lesson plan

Reactivate knowledge

1 Name the member of the electromagnetic spectrum with the lowest frequency.

2 Which part of the electromagnetic spectrum can human eyes detect?

3 What is the name of the part of the electromagnetic spectrum that lies between visible and microwave?

Answers: *1* Radio waves *2* Visible light *3* Infrared

Common misconception

Students may think that infrared and ultraviolet light are visible to the human eye, because infrared light can be seen from very hot glowing objects and ultraviolet light can be seen under special lamps.

Trigger interest

Colour filters (10 mins)

Shine a bright white light through a series of filters and ask students to choose, with the use of mini-whiteboards, how they would like to explain what is happening – either with words or with a diagram. Be aware that students with colour vision deficiency may need additional support to interpret and describe what they observe. If this is the case, some groups of students could be given images showing how coloured filters absorb particular wavelengths of visible light, and they can be tasked to elaborate on these images. Following this activity, ask students to predict whether a similar effect would occur for the non-visible parts of the electromagnetic spectrum. This leads into the absorption of electromagnetic energy as it passes through materials, which the students will explore in the required practical.

Exposition

Uses of electromagnetic radiation (20 mins)

Wherever possible, use mini-whiteboards to draw out students' wider knowledge of the applications of different parts of the electromagnetic spectrum. For example, before going into detail, students could guess how each wave is used in different contexts from visual clues or prompts provided.

- Infrared radiation – emphasise the relatively low energy carried by the waves but explain that high intensity (related to brightness) can mean that lots of energy is delivered by electric heaters, and so on. In this way, overexposure to infrared radiation can lead to skin damage and burning.
- Microwaves – describe uses of microwaves, for example, phones and microwave ovens. Specify that water and fat molecules in foods are able to absorb microwaves, and this ability to absorb microwaves produces a heating effect. Extend this example to explore why microwave ovens can cook food but mobile phones (another emitter of microwaves) do not – drawing out ideas about intensity.
- Radio waves – radios and games consoles (i.e., bluetooth controllers) are good examples, and emphasise that these signals can penetrate walls. Extend this example to discuss absorption of radio waves in metal wires as a part-explanation of how antennae work.

Main activity

Radiation and absorption (20 mins)

Students investigate the absorption and emission of infrared radiation in the required practical and then answer the questions. If students are planning their own investigation, ensure they identify the independent, dependent, and control variables correctly before starting. The experiment takes some time, especially if looking at absorption. In this case, it might be more time-efficient to set up before the lesson's exposition so that the cans have 20 minutes or so to warm before the students begin analysis.

Review and practice

What's the frequency? (5 mins)

Students carry out an interactive activity, where they calculate wavelengths of some radio stations when given the frequency. Students should then rearrange the equation to calculate the radio station frequency, when given the wavelength.

P13.3 Communications

AQA GCSE Physics specification points

4.6.2.2 Properties of electromagnetic waves 1

4.6.2.3 Properties of electromagnetic waves 2

4.6.2.4 Uses and applications of electromagnetic waves

Smart Curriculum Narrative

Waves

Theme/subtheme

Waves/Electromagnetic waves

Disciplinary knowledge

- MS 1c, 3b, 3c

Practical

Optical fibres for communication

Learning objectives	Learning outcomes: Developing Aiming for 3	Learning outcomes: Secure Aiming for 5	Learning outcomes: Extending Aiming for 8
Describe how electromagnetic waves are used in communication	State that radio waves and microwaves are used in communications	**Describe how electromagnetic waves are used in communication**	(H) Explain why certain wavelengths are suited to their application
(H) Describe how radio waves are produced and detected	State that an aerial is needed to detect radio signals	**(H) Describe how radio waves are produced and detected**	Explain how radio waves are detected
Analyse data relating to the risks of using radiation for communication and draw a conclusion	Identify possible risks of using radiation from provided data	**Analyse data relating to the risks of using radiation for communication and draw a conclusion**	Justify a conclusion about the safety of using radiation for communication

Tier 2 vocabulary	Tier 3 vocabulary
digital, light, radio wave, *rate*, speed	aerial, astronomy, electromagnetic spectrum, extraterrestrial, frequency, *microwave*, optical fibre, signal, transmission, transmit, *vacuum*, wavelength

Lesson plan

Reactivate knowledge

1 Name a wavelength used in fibre optic communications.

2 What does the prefix mega mean?

3 Which member of the electromagnetic spectrum is used to communicate with satellites?

Answers: 1 Visible light or infrared *2* Million or x10^6 *3* Microwave

Common misconception

Students may think that radio waves travel at the speed of sound. Tell students that radio waves are not sound waves. They are part of the electromagnetic spectrum, which is a set of waves that all travel at the same speed through a vacuum, which is 3×10^8 m/s. This is the speed of light. All waves on the electromagnetic spectrum, in this sense, only differ by their frequency and wavelength, so can all be considered as 'light' waves.

Trigger interest

Wow! (5 mins)

The 'wow signal' was a radio wave that originated in space and was detected by astronomers on Earth in 1977. Students should carry out independent research to find one or two additional facts about the wow signal, writing them on sticky notes posted to a wall/whiteboard.

Read a selection of student statements back to the class, emphasising the position of radio waves in the electromagnetic spectrum and their use in the field of astronomy, and especially in the search for extraterrestrial intelligence.

Exposition

Radio waves and mobile phones (20 mins)

Discuss the various regions of the radio spectrum, with a particular emphasis on the position of microwaves within it. Ensure that the link between rate of data transfer and frequency is explicit – microwaves provide the greatest rate of information transfer.

Discuss how signals are transmitted and the need for transmission towers. Images are useful to illustrate the route of information between phones and the towers (i.e., mobile phones do not communicate directly). An image or demonstration of an older mobile phone may be useful to show students the aerial.

Present an article headline that speculates on the link between mobile phone use and brain damage. Ask students for their opinions based on the scientific evidence they may have come across. Explain that mobile phone signals are not likely to have negative effects on the brain because the intensity is low.

Main activity

Optical fibres for communication (20 mins)

Demonstrate a simple transmission of light from a torch pulsed through a fibre cable with a convoluted path. Remind students that visible light and infrared waves have a much higher frequency than radio waves and so fibres will transmit data at a higher rate than radio signals. Students can compare energy and information transfer through a fibre cable with that in radio transmissions.

Higher tier candidates also need to discuss signal and carrier waves. The process here is quite complex, so lead students through each stage carefully. Foundation tier candidates can spend this time researching health and safety precautions when using mobile phones.

Review and practice

The benefits of digital inclusion (10 mins)

Ask students to discuss and suggest the benefits of optical fibre communications, for example, for internet access. Explore the issue of digital inclusion and exclusion, using examples of large-scale investment by local or international government to establish reliable internet access, and the social and educational benefits of this technology.

P13.4 Ultraviolet waves, X-rays, and gamma rays

AQA GCSE Physics specification points

4.6.2.3 Properties of electromagnetic waves 2

4.6.2.4 Uses and applications of electromagnetic waves

Smart Curriculum Narrative

Waves

Theme/subtheme

Waves/Electromagnetic waves

Disciplinary knowledge

- WS 1.5
- MS 1b

Activity

Ultraviolet waves

	Learning outcomes		
Learning objectives	**Developing Aiming for 3**	**Secure Aiming for 5**	**Extending Aiming for 8**
Describe the ionisation of atoms in simple terms	State that high-frequency electromagnetic radiation is ionising	**Describe the ionisation of atoms in simple terms**	Describe in detail the effect of ionising radiation on living cells
Describe the dangers of ultraviolet radiation	Describe the uses of ultraviolet (UV) radiation	**Describe the dangers of ultraviolet radiation**	Compare different regions of the electromagnetic spectrum in terms of their potential harmfulness
Describe the similarities and differences between X-rays and gamma radiation	Describe the uses and dangers of X-rays and gamma radiation	**Describe the similarities and differences between X-rays and gamma radiation**	(H) Explain why gamma radiation is suited to its application

Tier 2 vocabulary	Tier 3 vocabulary
atmosphere, atomic, DNA, electron, energy, *incident*, light, ozone, visible, wave radiation	absorb, electromagnetic, gamma radiation, imagery, infrared, ionisation, stellar, ultraviolet, wavelength, X-ray

Lesson plan

Reactivate knowledge

1 Which member of the electromagnetic spectrum has the highest frequency?

2 Which has the longer wavelength: light or ultraviolet?

3 What is meant by ionisation?

Answers: 1 Gamma *2* Light *3* When an atom loses an electron/ becomes an ion

Common misconception

Students may think that having an X-ray photograph causes cancer. Tell students that although ionising radiation like X-rays can lead to cancer if somebody is exposed to a large amount of it, and for a sustained period of time, a medical X-ray photograph exposes a patient to only a minimal amount of radiation. Medical experts keep an eye on how much radiation exposure a patient receives and ensures it is well below dangerous levels.

Trigger interest

Stellar imagery (10 mins)

Show a range of images of the Sun in different wavelengths (visible light, ultraviolet, infrared etc) and ask students to discuss and propose what is common in each image and what is different. The same activity could be carried out for different planets in the Solar System, for example, comparing images of Jupiter.

Discuss how these images are captured, why they look different, and which parts of electromagnetic radiation reach the surface of the Earth.

Extend this idea by showing a graph of the transmission characteristics of the Earth's atmosphere.

Exposition

Ultraviolet , X-rays, and gamma rays (25 mins)

Start by demonstrating the effect of ultraviolet light on a range of objects, as outlined in the practical demonstration. This is to emphasise that ultraviolet light really exists – because we can indirectly observe the effects it has on objects that are illuminated by it. Emphasise the damage it can cause, particularly to the eyes and skin, because of the higher energy carried by this part of the electromagnetic spectrum.

Darker skin tones can produce more melanin than lighter skin tones. Melanin is capable of blocking a certain amount of ultraviolet radiation from the sun, so that a person with more melanin is less likely to get sunburned. This is an evolutionary response to environment and climate. However, it does not make a person immune to the harmful effects of ultraviolet radiation, increasing the risk of skin cancer without causing sunburn.

Move to even higher frequency/energy waves, emphasising the penetrating power through different materials. A few simple non-medical X-ray images can be used to outline a use for X-rays.

Finally, outline the key uses of gamma radiation; students should deduce that this is the highest frequency/energy region of the electromagnetic spectrum, and therefore that gamma rays have the highest penetrating power. Link this high energy to ionisation effects and damage to living cells. The process of ionisation should be visually explored, showing students the basic process of atomic electron removal. This enables them to link it to a real, physical change at the atomic level, and therefore, alteration (and in most cases, damage) to DNA.

Students should understand that greater and longer exposure leads to a higher risk of damage. They can then suggest various control procedures, for example, reducing exposure time, using protective clothing, increasing the distance from the source, and dose measurement.

Main activity

Ultraviolet waves (10 mins)

Demonstrate the effect of ultraviolet light on different clothing whilst explaining that they absorb the ultraviolet light and then emit visible light.

This demonstration can be extended by using additional objects. For example, students will notice a different effect of ultraviolet light incident on plastic money, i.e., a £5 note, compared to that of plain white paper.

Review and practice

One world (10 mins)

Interactive activity in which students complete a paragraph to briefly discuss the problem with the ozone layer that was discovered in the 1980s and the steps taken to tackle it, to show that global issues can be solved when international governments work together.

Extend this activity by circulating and asking students to suggest other environmental issues that could be addressed with the cooperation of government and international bodies.

P13.5 X-rays in medicine

AQA GCSE Physics specification points

4.6.2.3 Properties of electromagnetic waves 2

4.6.2.4 Uses and applications of electromagnetic waves

Smart Curriculum Narrative

Waves

Theme/subtheme

Waves/Electromagnetic waves

Disciplinary knowledge

- WS 1.5
- MS 1a

Activity

Case studies

Learning objectives	Learning outcomes: Developing – Aiming for 3	Learning outcomes: Secure – Aiming for 5	Learning outcomes: Extending – Aiming for 8
Describe the factors that affect the radiation doses received by people	State that radiation dose is the measure of the risk of harm when exposed to radiation	**Describe the factors that affect the radiation doses received by people**	Evaluate the doses of ionising radiation received by analysing data
H Explain why X-rays are suitable for destroying cancer cells	State that X-ray therapy can be used to destroy cancerous cells in the body	**H Explain why X-rays are suitable for destroying cancer cells**	Explain the precautions taken to protect healthy tissue when using X-rays to destroy cancerous cells
Describe the formation of an X-ray photograph in terms of absorption or transmission	State that X-rays are absorbed more by bones and teeth than soft tissue	**Describe the formation of an X-ray photograph in terms of absorption or transmission**	Explain why contrast media can be used during X-rays

Tier 2 vocabulary	Tier 3 vocabulary
anatomy, atom, dynamo, electricity, radiation	absorption, CT scanner, ionisation, radiation, transmission, X-ray

Lesson plan

Reactivate knowledge

1 Which is more dangerous inside the body: alpha, beta, or gamma radiation?

2 What metal can absorb X-rays and gamma rays?

3 Which member of the electromagnetic spectrum has the most energy?

Answers: 1 Alpha (as it is the most ionising) 2 Lead 3 Gamma

Common misconception

Students may think that X-rays cause objects to become radioactive. Tell students that during a medical X-ray photograph, X-rays pass through the body and do not get trapped in it. Therefore, humans cannot expose other humans to radiation in this way. 'Radioactive' atoms give off radiation because they have excess energy – and radioactive objects contain a lot of radioactive atoms. Sometimes, humans and other objects can become contaminated with radioactive atoms, for example, by coming into contact with a radioactive substance, or even by eating certain foods!

Trigger interest

***Hand mit ringen* (5 mins)**

Show students the very first medical X-ray image and ask them what it shows and how it might have been made. It is an image of Wilhelm Röntgen's wife's hand, wearing rings.

Exposition

Medical X-rays and CT scanners (15 mins)

Outline the process of a medical X-ray photograph, supporting this with a range of real images. Discuss and label the images in terms of absorption and transmission of the X-rays, including images that use contrast media to differentiate tissues.

Remind students of the procedures used to limit radiation exposure; they should recognise the unit sievert and use it in comparisons, although they do not need to remember it.

Main activity

Case studies (25 mins)

Provide students with resources to investigate the use of a CT scanner. This is a very good opportunity to discuss and compare doses; CT scans typically give up to 9 mSv (three years worth of normal background exposure) compared to roughly 0.1 mSv for a chest X-ray. Students should present their findings in a table, showing similarities and differences between X-rays and CT scans. Higher tier candidates should include a risk/benefit analysis, with descriptions of how CT scans are used to detect cancerous cells while X-rays are used to treat them.

Be mindful of any student who has personal experience with cancer treatment – for example, an appropriate disclaimer might be helpful at the start of this activity.

Review and practice

X-ray cars (10 mins)

Give students a short historical account of the groundbreaking scientific work of Marie Curie on the medical use of X-rays; the first woman to win a Nobel prize and the first person to win twice.

During World War One, she developed mobile X-ray units, known as 'X-ray cars', a vehicle containing an X-ray machine and photographic darkroom equipment – which could be driven right up to the battlefield where army surgeons could use them to guide their surgeries. Curie also developed the use of a dynamo to generate electricity in the vehicles and trained many other women on the physics of electricity and X-rays. She also provided practical lessons on anatomy and photographic processing.

P13.6 Infrared radiation in more detail

AQA GCSE Physics specification points

4.6.3.1 Emission and absorption of infrared radiation

4.6.3.2 Perfect black bodies and radiation

Smart Curriculum Narrative

Energy

Theme/subtheme

Energy/Energy and temperature

Disciplinary knowledge

- MS 4a, 4c

Practical

Black body radiation

	Learning outcomes		
Learning objectives	**Developing Aiming for 3**	**Secure Aiming for 5**	**Extending Aiming for 8**
Describe how the rate of emission of infrared radiation is related to the temperature of a body	State that all objects emit and absorb infrared radiation	**Describe how the rate of emission of infrared radiation is related to the temperature of a body**	Explain how thermal cameras can be used to detect objects
Explain how the intensity and wavelength of radiation emitted depends on the temperature of the object	State what is meant by a black body	**Explain how the intensity and wavelength of radiation emitted depends on the temperature of the object**	Compare the black body spectra of two objects to identify which is at a higher temperature
Describe the visible changes in an object's emitted radiation as its temperature is increased	State that the emitted radiation depends on the temperature of an object	**Describe the visible changes in an object's emitted radiation as its temperature is increased**	Explain the visible changes in an object's emitted radiation as its temperature is increased

Tier 2 vocabulary	Tier 3 vocabulary
absorb, absorber, Bunsen burner, electric, emit, emitter, filament lamp, intensity, magnetic, microwave, prism, radiation, radio wave, spectrum, star, Sun, temperature, thermometer, transmit, ultraviolet ray, voltage, X-ray	black body, black body radiation, electromagnetic radiation, electromagnetic spectrum, electromagnetic wave, infrared radiation, potential difference, wavelength

Lesson plan

Reactivate knowledge

1 Name the colours in the white light spectrum.

2 Which type of surface is the best at absorbing infrared radiation?

3 Which type of surface is the best at reflecting infrared radiation?

Answers: 1 Red, orange, yellow, green, blue, indigo, violet *2* Dark, matte surfaces *3* Light, shiny surfaces

Common misconception

Students may think that infrared radiation is harmful. Respond that it is not harmful and that it's a naturally occurring form of radiation that travels to Earth from the Sun and is absorbed by objects on Earth as well as the Earth's surface itself – causing a rise in temperature. All objects – even human beings – emit radiation, and most of the radiation we emit is in the infrared region of the electromagnetic spectrum.

Trigger interest

Seeing at night (5 mins)

Gamers may already be familiar with infrared goggles. Show students infrared images of living objects (e.g., humans, animals) and ask them to identify them. Guide students to realise hotter regions are emitting more infrared radiation. A labelled diagram of the electromagnetic spectrum helps to reinforce the sameness of visible light to infrared light, except in wavelength.

Extend this discussion to explore why other living things, like plants, are not as 'bright' when viewed with infrared goggles compared to other things, like humans.

Exposition

The bright Sun (5 mins)

Using an image of the Sun, emphasise that when people are in sunlight, they are absorbing infrared radiation, which causes their temperature to rise.

Link infrared waves back to the electromagnetic spectrum – they are very similar to visible red light waves, but they have a much longer wavelength. Then launch the practical activities to show how infrared radiation can be detected.

Main activity

Detecting infrared radiation (15 mins)

In a very dark room, carry out the demonstration shown in Figure 1 in the Student Book. Ask a student to note the temperature rise on the thermometer, and to describe to the class what happens to the temperature as they move the thermometer away from the screen.

Black body radiation (20 mins)

Emphasise that all objects emit infrared radiation, and that hotter objects emit more infrared radiation in a given time. This practical guides students to gradually increase the potential difference across a low-voltage filament lamp, up to its operating value.

A very dark room is required. Build up the 'wonder' by asking students to carefully describe what they see as the potential difference across the lamp increases. When the potential difference is very low, students will correctly say they see nothing – draw out what they might see if they were wearing infrared goggles (linking back to the starter). As the potential difference gradually increases, they should see that the lamp firstly glows a dull red colour, then becomes white and brighter as it gets hotter (an increase in total radiation emitted). This is a good opportunity to link directly to the electromagnetic spectrum, emphasising the position of infrared relative to the visible part of the spectrum.

Review and practice

Black body radiation (10 mins)

Revisit the image of the Sun. Introduce the definition of a black body, emphasising that the Sun (and other stars in our universe) are almost perfect black bodies. Discuss the radiation curve that a perfect black body produces – it is important that students recognise that the shape of this curve depends on the temperature of the body (a higher temperature shifts the most intense wavelengths to being shorter).

Sketch the radiation curve next to diagrams/images of the lamps from the black body radiation practical at each stage of varying potential difference. Students should immediately conclude that the shift in the position of the maximum wavelength (toward shorter wavelengths) corresponds to the temperature increasing.

Extend these ideas by linking black body curves to analysis of light from stars to determine and compare their surface temperature.

P13.7 Infrared radiation and temperature

AQA GCSE Physics specification points

4.6.3.2 Perfect black bodies and radiation

Smart Curriculum Narrative

Energy

Theme/subtheme

Energy/Energy and temperature

Disciplinary knowledge

- WS 1.2
- MS 3a

Activity

Greenhouse effect/Cause and effect

	Learning outcomes		
Learning objectives	**Developing Aiming for 3**	**Secure Aiming for 5**	**Extending Aiming for 8**
Describe the heating of objects in terms of the rate of emission and absorption of radiation	State that all objects are continually emitting and absorbing radiation	**Describe the heating of objects in terms of the rate of emission and absorption of radiation**	Apply the concepts of absorption and emission of radiation to explain why an object maintains a constant temperature
Describe the factors affecting the temperature of the Earth	State that radiation from space can be reflected, absorbed, or emitted by the Earth	**Describe the factors affecting the temperature of the Earth**	Describe the relationship between the incoming radiation and the emitted radiation, if the Earth's temperature is to stay constant
Explain how the Earth's temperature is affected by the absorption and emission of radiation	Describe the greenhouse effect in terms of absorption and emission of radiation	**Explain how the Earth's temperature is affected by the absorption and emission of radiation**	Draw and interpret diagrams to explain how radiation affects the temperature of the Earth

Tier 2 vocabulary	Tier 3 vocabulary
absorb, absorption, atmosphere, carbon dioxide, degree Celsius, Earth, emission, emit, energy, methane, Moon, radiation, reflect, Space, Sun, temperature, vapour	greenhouse gas, infrared radiation, wavelength

Lesson plan

Reactivate knowledge

1 What is infrared radiation?

2 What objects emit infrared radiation?

3 How can you detect infrared radiation?

Answers: 1 Part of the electromagnetic spectrum with longer wavelength than visible light *2* All objects *3* Thermal cameras/felt as heat by the skin

Common misconception

Students may think that ultraviolet rays only damages light skin. Tell students that it is true that people with lighter skin are at a higher risk of developing certain types of skin cancer. However, skin cancer can affect anyone regardless of age, race, gender, or ethnicity. The relationship between melanin (the pigment responsible for different skin tones) and ultraviolet radiation is complex.

Trigger interest

Warm words (10 mins)

Show students the famous 'hockey stick' graphs that show drastic global temperature increases over the past 50 years and projections of future increases. There is plenty of opportunity to link this lesson to real, current events (e.g., record summer temperatures in the UK and Europe). Use the interactive to suggest and discuss possible causes of this heating (e.g., human actions and natural effects, like El Niño). Introduce a graph showing changes in the CO_2 concentration of the atmosphere for further discussion.

Exposition

Radiation and surfaces (10 mins)

Recap ideas about the emission and absorption of infrared radiation as discussed in the previous lesson, including examples of shiny and dark surfaces. Support students by providing plenty of visuals to link to real-life examples (e.g., using kitchen foil to keep food warm).

Next, explore human skin as an example of a surface that emits and absorbs infrared radiation. Discuss the effect of material and colour used in clothing and protective gear (e.g., cold-weather clothing, anoraks, and emergency blankets).

Main activity

Greenhouse effect (10 mins)

Introduce the Earth's surface as one that emits and absorbs infrared radiation. Make explicit the analogy that greenhouse gases in the atmosphere act as a covering that regulates the temperature of the Earth over an extended period – much like clothing or kitchen foil. Emphasise the difference in wavelength of radiation emitted by the Sun and that emitted by the Earth, linking this to the different temperatures of each body.

Introduce the idea that greenhouse gases trap infrared radiation emitted by the Earth, to prepare students to answer Question 3 in the Student Book about the positive aspects of the greenhouse effect.

Cause and effect (10 mins)

Students should explore evidence for the increase in greenhouse gas concentration and be able to list a range of effects, including increasing average temperatures and rising sea levels.

Enrich discussions by prompting students to think about whether these causes are natural or result from economic/geopolitical activity. This is a great opportunity to emphasise that higher average temperatures and climate change will have varying effects around the world, on people as well as nature, and the reality of how these effects may impact some nations and groups of people disproportionately.

Review and practice

Cut and dry? (15 mins)

Discuss global warming and it's potentially drastic consequences for nature and humans. This is an opportunity to talk about the nature of science as a constantly evolving body of knowledge that requires global perspectives and verifiable evidence. Explore social, economic, and political reasons as to why some countries, communities, and individuals may doubt or reject the notion that humans are negatively impacting the climate.

Extend discussions by introducing case studies of national and international targets for reaching net-zero emission and the implications of irreversible climate effects. Balance the discussion by providing positive notes of hope; for example, boosts for clean energy legislation around the world, significantly falling rates of deforestation in the Amazon rainforest, and the creation of fund relief for countries most vulnerable to climate change.

P13 Electromagnetic waves

Introduction to checkpoint intervention

The checkpoint intervention provides suggestions for a lesson to follow up the *P13 Electromagnetic waves* checkpoint assessment. Depending on the outcomes of the assessment, these suggestions can be used to move a class towards secure, to target specific learning outcomes, or to consolidate knowledge for students achieving secure outcomes. Use students' outcomes from the checkpoint assessment to address general misconceptions from the content covered. Be prepared to re-cover content as required; three key concepts from the chapter are covered in more detail below.

Checkpoint secure learning outcomes

- Describe how light, infrared, microwaves, and radio waves are used in a variety of scenarios.
- Plan an investigation into the rate of cooling due to emission of infrared radiation.
- Describe the similarities and differences between X-rays and gamma radiation.
- Describe the ionisation of atoms in simple terms.
- Describe the formation of an X-ray photograph in terms of absorption or transmission.
- Calculate the frequency and the wavelength of an electromagnetic wave.

Student reflection

Encourage students to reflect on whether there were any checkpoint questions they found difficult or straightforward as well as their preparation for the checkpoint assessment, including how they revised, what they revised, the time spent on revision, and what areas they could improve upon. For more strategies, see the Metacognition spread.

Intervention activity: Describe how light, infrared, microwaves, and radio waves are used in a variety of scenarios.

For intervention activities on electromagnetic waves, wherever possible, have an annotated diagram to hand showing the different parts of the electromagnetic spectrum, including wavelength and frequency labels. Students should recognise the visible part of the spectrum, and the red 'end' (emphasise, near infrared) and the violet 'end' (emphasise, near ultraviolet). For students with colour blindness or other visual impairments, ensure they can match colours, when appropriate, using felt pens or other comparative methods.

Print a selection of images that demonstrate uses of light, infrared, microwaves, and radio waves in a variety of simple scenarios. Print a selection of sentences that describe the use, for example, microwaves heat food, and radio waves communicate via satellite. Students should match the sentences with the images and discuss their conclusions. Develop this activity by asking for additional examples of uses that students may recall from prior knowledge. Encourage students to find these examples independently and create their own image/sentence matches.

Intervention activity: Describe the ionisation of atoms in simple terms.

Be aware of the cognitive load of learning, and understanding, multiple keywords and processes in this intervention activity. Using a simple image or model of an atom, ask students to label the nucleus and the electrons.

Develop this by asking students to identify the particles in the nucleus (protons and neutrons), and the charges of these and of the electrons (positive, neutral, and negative). If appropriate, have labels/keywords available so students can select/drag and drop.

It is important that students are able to conclude that atoms are neutrally charged, because the number of protons (positive charges) balances the number of electrons (negative charges).

Demonstrate how atoms turn into ions by adding/removing electrons. Students should then be able to distinguish atoms from ions by comparing numbers of protons to electrons.

In simple terms, ionisation is the removal of an electron from an atom to form a charged ion. In fuller definitions, it is the removal or capture of an electron to form a charged ion, but for the purposes of this intervention, focus on the removal process. Emphasise that removing an electron requires energy; if an electron gets enough energy (say, from an incoming X-ray), then it can escape its atom. This is called ionisation, because the atom turns into an ion. A video or animation showing this process is effective.

Intervention activity: Calculate the frequency and the wavelength of an electromagnetic wave.

Present students with the hinge question: *How fast do all light waves travel?*

They might already know that the speed of light is special, and that it is the maximum universal speed. Using mini-whiteboards, explore their knowledge of the speed of light, leading to the key fact that all electromagnetic waves (including visible light) travel at the same speed in a vacuum, 3×10^8 m/s.

It is important to emphasise that these are waves, so that students can link this speed to the wave speed equation.

There are a variety of equation frameworks to help students use and rearrange equations. Equation triangles tend to be useful interventions to build student confidence with using equations. Model how this is used for the wave speed equation. Show explicit substitution of the speed of light, 3×10^8 m/s, and how to rearrange the equation for frequency and wavelength. Ensure that a variety of examples are used, for example, infrared, microwaves, visible light, and gamma rays. Students should recognise, in each case, that the wave speed remains constant.

P14 Light

P14.1 Reflection of light

AQA GCSE Physics specification points
4.6.1.3 Reflection of waves

4.6.2.6 Visible light

Smart Curriculum Narrative
Waves

Theme/subtheme
Waves/Wave properties

Disciplinary knowledge
- WS 1.2
- MS 5a, 5b, 5c

Practical
Mirror mirror

	Learning outcomes		
Learning objectives	**Developing Aiming for 3**	**Secure Aiming for 5**	**Extending Aiming for 8**
Construct accurate ray diagrams showing the reflection of light rays	State the law of reflection	**Construct accurate ray diagrams showing the reflection of light rays**	Draw a ray diagram showing the position of an image in a plane mirror
Define real and virtual images	Investigate the reflection of light from different surfaces	**Define real and virtual images**	Explain the difference between real and virtual images
Draw diagrams to show reflection from smooth and rough surfaces	State that reflection from a smooth surface is called specular reflection and from a rough surface is called diffuse reflection	**Draw diagrams to show reflection from smooth and rough surfaces**	Explain why some surfaces form images during reflection but others do not

Tier 2 vocabulary	Tier 3 vocabulary
boundary, light, *media*, pressure, protractor, *ray*	image, incidence, incident, reflection, refraction, *virtual*

Lesson plan

Reactivate knowledge

1 State the law of reflection.

2 What type of image is formed in a mirror?

3 Compare the size of the image of an object in a plane mirror to the object.

Answers: 1 Angle of incidence = angle of reflection *2* Virtual *3* Same size

Common misconception

Students may think that light is only reflected from shiny surfaces. Respond that reflection refers to a change of direction when a wave, including light, meets a boundary between two different media. The type of surface at the boundary (e.g., shiny, rough, smooth, etc.) determines *how* the light changes direction.

Trigger interest

Lunar see (10 mins)

Interactive activity where students complete an explanation of how we are able to see the Moon.

Use the hinge question: *What natural phenomenon (an event in nature that occurs without human input) allows us to see the Moon?*

Students could, using one word only, vote on mini-whiteboards before attempting the interactive.

Support this activity by providing a list of phenomena that students could select from, including a 'not sure' option (e.g., reflection, refraction, heating, pressure, etc.)

Exposition

Ray of light (15 mins)

Check students' ability to recognise a 'ray of light', and then to construct a ray diagram that shows reflection of a ray of light. For instance, you could show four diagrams showing an incident ray undergoing various direction changes, with only one showing a correct representation of reflection.

Build the concept of reflection from this basis, ensuring students correctly understand the rule of reflection. Producing a ray with a ray box to demonstrate reflection at a plane mirror, and checking with a protractor (e.g., under a visualiser), is a nice demonstration to supplement this theory (or, equivalently, using a PhET simulation).

Outline real and virtual images, contrasting the image produced by a projector to an image produced by a mirror. Ensure that you look at pairs of rays to allow the image to be located properly, e.g., Figure 3 in the Student Book. Extend this concept by showing how (virtual) rays are constructed behind the mirror to form the virtual image.

Main activity

Mirror mirror (25 mins)

Students observe their images in plane mirrors and describe their properties. Allow students to investigate a range of surfaces that produce clear images, blurry images, and then no image at all, ensuring they are given opportunity to articulate what they see. Surfaces should be described in terms of roughness or smoothness.

Review and practice

Pepper's Ghost (5 mins)

Show students a video online of Pepper's Ghost, the classic optical illusion, or recreate it as a demonstration. The idea of partial reflection can then be discussed; students can recreate observations of the Pepper's Ghost effect by looking through a window and describing simultaneous observations of what is outside, and their own reflections.

Revisit the hinge question; students should identify correctly that reflection is the natural phenomenon that allows the Moon to be observed, and link this to the Pepper's Ghost effect.

P14.2 Refraction of light

AQA GCSE Physics specification points

4.6.1.3 Reflection of waves

4.6.2.2 Properties of electromagnetic waves 1

Smart Curriculum Narrative

Waves

Theme/subtheme

Waves/Light

Disciplinary knowledge

- WS 1.2
- MS 1a, 5a, 5c

Required Practical 9

Light

	Learning outcomes		
Learning objectives	**Developing Aiming for 3**	**Secure Aiming for 5**	**Extending Aiming for 8**
Construct accurate ray diagrams showing the refraction of light rays	Draw a normal at the point where light enters a glass block	**Construct accurate ray diagrams showing the refraction of light rays**	Construct accurate ray diagrams to explain apparent depth
Investigate the refraction of light through a glass or Perspex block	Measure the angle of incidence and angle of refraction for a simple refraction	**Investigate the refraction of light through a glass or Perspex block**	Analyse the data from a refraction investigation to test different substances to determine whether it fits a suggested relationship

Tier 2 vocabulary	Tier 3 vocabulary
boundary, density, light, *media*, *medium*, speed	air-glass boundary, dispersion, incidence, incident, prism, reflect, refract, refraction

Lesson plan

Reactivate knowledge

1 What is refraction?

2 Why does refraction happen?

3 What is a normal on a ray diagram?

Answers: 1 Change in speed of wave at a boundary between two media
2 The media have different densities *3* A line drawn at 90° to the surface where the incident ray meets the surface

Common misconception

Students may think that refraction is the change in direction when light meets a boundary between two media. Tell students that refraction refers to the change in *speed* of a wave – including a light wave – when it meets a boundary between two media. Waves will either speed up, or slow down, if the media have different densities.

In most cases, this change in speed is accompanied by a change in direction, but not in all cases. Sometimes, if the light is incident at 90° to the boundary, its speed still changes but its direction does not.

Trigger interest

Stealth beaker (5 mins)

Place one small glass beaker inside another. Slowly fill the inner beaker with transparent baby oil and then fill the outer beaker. The inner beaker becomes invisible.

As a link back to P14.1, ask students the hinge question: *What natural phenomenon causes the inner beaker to disappear?*

As before, gather a selection of votes using mini-whiteboards, and support their suggestions, where appropriate, using a bank of keywords.

Exposition

Key terms (15 mins)

Take care when using keywords and phrases while delivering this topic; it should not be assumed that these are all immediately clear to students. Ensure terms such as: medium, media, density, boundary, incidence, dispersion, and the convention for describing a boundary, are questioned or explained (e.g., 'air-glass' means that light is incident in air and crosses the boundary into glass). Ask what would a glass-air boundary mean, to check understanding.

Introduce the definition of refraction, highlighting the common misconception about changes in direction. Emphasise that, in most cases, a change in direction is observed. Show this using available equipment (or PhET simulation) of a light ray refracting at an air-glass boundary, identifying the angles clearly (using a visualiser, if available). Link this to the behaviour of waves studied in topic P12.3, and support this further by turning the demonstration into a large, labelled ray diagram.

Then, show the dispersion of light through a prism, noting which colours are refracted the most. Link this to a larger change in speed for that colour (wavelength).

Main activity

Light (25 mins)

Students investigate the refraction of light at boundaries, noting that the change of direction increases when the angle of incidence does.

They should then plan an investigation to explore different materials – to show that the difference between materials is important. This can be carried out in the next lesson.

Some students may be interested in the mathematical relationship between the angle of incidence and the angle of refraction; the challenge could be to verify the formula for this relationship, $\frac{\sin i}{\sin r} = \text{constant}$ using their experimental data.

Review and practice

Reflect or refract? (10 mins)

This interactive activity extends students' abilities to distinguish between reflection and refraction. Students complete an explanation on the difference between reflection and refraction, using their new understanding from this topic. They then sort a series of scenarios, identifying whether they show reflection or refraction.

Circulate, and assess students' ideas, revisiting the hinge question with small groups. A sketch, showing how apparent invisibility is accomplished by refracting light away from our eyes, is a useful prompt to support their conclusions.

P14.3 Light and colour

AQA GCSE Physics specification points

4.6.1.3 Reflection of waves

4.6.2.2 Properties of electromagnetic waves 1

4.6.2.6 Visible light

Smart Curriculum Narrative

Waves

Theme/subtheme

Waves/Light

Disciplinary knowledge

- WS 1.2
- MS 1a, 5a, 5c

Required Practical 9

Light

Learning objectives	Learning outcomes: Developing Aiming for 3	Learning outcomes: Secure Aiming for 5	Learning outcomes: Extending Aiming for 8
Explain the effect on the different wavelengths of visible light as they pass through a filter	Describe the difference between translucent and transparent	**Explain the effect on the different wavelengths of visible light as they pass through a filter**	Explain what colour would be observed if a white light was observed through a red and a blue filter at the same time
Describe how objects reflect different wavelengths of visible light to appear different colours (including white or black)	State that coloured objects reflect the colour they appear to be	**Describe how objects reflect different wavelengths of visible light to appear different colours (including white or black)**	Explain how the colour of an opaque object is related to how different wavelengths of visible light are reflected or absorbed
Determine the apparent colour of a coloured surface when illuminated by different colours of light	State that colours appear different when observed through a colour filter	**Determine the apparent colour of a coloured surface when illuminated by different colours of light**	Explain the apparent colour of a coloured surface when illuminated by different colours of light

Tier 2 vocabulary	Tier 3 vocabulary
light, spectrum	absorb, dispersion, opaque, prism, reflection, translucent, transmit, transparent, wavelength

Lesson plan

Reactivate knowledge

1 Name the colours of the spectrum in order of increasing wavelength.
2 How can you separate the colours of white light?
3 What can happen when light strikes a material?

Answers: 1 Violet, indigo, blue, green, yellow, orange, red *2* Refraction/using a prism *3* Reflected, refracted, transmitted, absorbed

Common misconception

Students may think that colour is a property of an object and is independent of the illuminating light and the receiver (the eye). Respond that it is *not* an intrinsic property of an object. Rather, objects absorb and transmit light in specific ways. The colour we observe tells us what light is absorbed by the object, and what light is transmitted. It is also important to consider what colour (wavelength) of light is originally illuminating the object. Most of the time, for casual observations, it is white light that illuminates objects. However, objects may appear as different colours if they are illuminated by different wavelengths of light.

Trigger interest

See through (10 mins)

Students use the interactive activity to match the keywords translucent, transparent, and opaque with definitions. They sort example materials into three groups, including some awkward materials that are barely transparent and some coloured filters.

Exposition

Note: Ensure that students with colour blindness, or other visual impairments, are supported appropriately. Consult with these students individually and encourage them to let you know whenever they think they may have issues with light or colour. It can be useful, for instance, to give students labelled coloured pens so that they can match these to other labelled coloured resources where needed, or to assign a 'colour buddy' to help them wherever needed.

Mixing colour (10 mins)

The mixing of light can be demonstrated using overlapping coloured light sources projected onto white surfaces. Demonstrate the mixing through different coloured filters. Intersperse demonstrations with reference to the spectrum of visible light, and narrow bands of wavelength and frequency for different colours. Explain how filters affect wavelengths in terms of absorption and transmission.

Figure 3 in the Student Book is a useful example of the types of diagram that can be used to assess students' understanding here.

Extend understanding of these observations, if appropriate, by asking students to predict effects of mixing light projected onto coloured surfaces.

Main activity

Light (30 mins)

Show the dispersion of white light by a prism by projecting it onto white paper. The students should briefly discuss light sources and the concept of a spectrum before moving on to complete this required practical to investigate reflection.

Review and practice

Disappearing colour (5 mins)

Students can observe the apparent production of a white surface during the rotation of a colour wheel. A demonstration or video works well – ask students to discuss their observations in pairs and suggest explanations for what they observe.

If there is time, or as an extension task, students can create their own 'Newton wheels', and provide a written explanation of the effect. Some students may be interested in the historical background as to why these devices are named after Isaac Newton.

P14.4 Lenses

AQA GCSE Physics specification points

4.6.2.5 Lenses

Smart Curriculum Narrative

Waves

Theme/subtheme

Waves/Light

Disciplinary knowledge

- WS 1.2
- MS 1c, 3b, 3c, 5a, 5b, 5c

Practical

Investigating the converging lens

Learning objectives	Learning outcomes: Developing Aiming for 3	Learning outcomes: Secure Aiming for 5	Learning outcomes: Extending Aiming for 8
Draw diagrams to show what happens to parallel rays of light incident on concave and convex, identifying the principal focus and focal length	Identify convex and concave lenses from their shapes	**Draw diagrams to show what happens to parallel rays of light incident on concave and convex, identifying the principal focus and focal length**	Explain ray paths through a lens in terms of refraction and the focal point
Calculate the magnification of a lens based on object and image size	State measurements of image height and object height must be in the same units when calculating magnification	**Calculate the magnification of a lens based on object and image size**	Perform calculations involving the rearrangement of the magnification equation
Describe sources of random and systematic error when investigating convex lenses	Identify sources of error in data when investigating convex lenses	**Describe sources of random and systematic error when investigating convex lenses**	Explain how to reduce random and systematic error when investigating convex lenses

Tier 2 vocabulary	Tier 3 vocabulary
light	concave, converge, convex, diverge, focal length, focal point, gravitational lensing, image, lens, plano-cylindrical lens, ray diagram, refract, *virtual*

Lesson plan

Reactivate knowledge

1 Does light speed up or slow down when it enters glass from the air?

2 How does the angle of incidence compare to angle of refraction when light enters glass from the air?

3 What is a virtual image?

Answers: 1 Slows down *2* Angle of incidence is greater than angle of refraction *3* One you cannot capture on a screen/paper

Common misconception

Students may think that when a lens is moved, an image will become bigger or smaller but will always remain sharp. Explain to students that lenses work by changing the direction of light as it passes through it and focusing the light (bringing all rays together) at a specific point. If the lens is moved, the point of focus will also change, so the image formed may be blurred.

Trigger interest

Sorted (5 mins)

Draw or display the shape of a convex and concave lens. Give students a pile of different lenses and ask them to identify and sort the lenses according to the shapes displayed.

They may independently research to find out the terms 'convex' and 'concave' themselves, or this can be given to them.

Exposition

The eye is a lens (10 mins)

Before introducing the idea that the eye is a natural lens, survey what students already know about lenses. For example, you could ask how many students wear glasses or contact lenses, or you could explore the role of opticians in understanding how lenses work.

Introduce the basic concept of a lens as an object that refracts light (loosely, changes the direction of light) as it passes through it. Show a diagram of converging and diverging rays, and an example of a focal point, and refer to an image of the human eye as containing a naturally occurring lens.

Extend this case reference, if appropriate, with a short comparison to a gravitational lensing effect. Show students an image of gravitational lensing as captured by the James Webb Space Telescope, emphasising the wonder of such images and how such grand effects confirm the principles of optics they are about to learn about in experiments. Ask students to guess what might be responsible for bending light on a galactic scale – which is the presence of huge amounts of mass (e.g., stars and galaxies). Just like a regular lens changes the path of light rays, so too can the gravitational fields of these large clusters of matter.

It is likely that students will have different prior knowledge and experience with lenses. Ensure plenty of time is given in this lesson for students to explore the effects of lenses themselves, with the practical equipment available. Exposition can happen as you circulate the room, scaffolding questioning as appropriate.

Main activity

Investigating the converging lens (35 mins)

Students investigate a converging lens; plano-cylindrical lenses can make this experiment simpler to set up.

This is an extensive practical that requires careful observations and manipulation of equipment, alongside activities that reinforce learning from their measurements. Ensure that roles are evenly distributed as appropriate, and activities are chunked so students have time to explore, discuss, and feedback their observations with you and/or each other.

Alongside the investigation, students should examine ray diagrams showing the ray paths through the lens, as these are crucial in later explanations. Ensure that you test whether or not students can identify a focal point, and describe the difference between a real and virtual image. Show students an example calculation of the magnification; they can use the Maths skills interactive for further practice.

Support can be provided by specifying the object distances and creating a ready-made results table. Provide partially completed ray diagrams and ask the students to complete them, for instance, by drawing the final ray paths.

Review and practice

Which lens? (5 mins)

Use the interactive activity to show students a range of diverging and converging lenses diagrams. Students identify which lenses are which and order them from longest to shortest focal length.

P14.5 Using lenses

AQA GCSE Physics specification points

4.6.2.5 Lenses

Smart Curriculum Narrative

Waves

Theme/subtheme

Waves/Light

Disciplinary knowledge

- WS 1.2
- MS 3b, 3c, 5a, 5b, 5c

Practical

Real image from a convex lens

	Learning outcomes		
Learning objectives	**Developing Aiming for 3**	**Secure Aiming for 5**	**Extending Aiming for 8**
Construct ray diagrams showing the formation of images by a convex lens when the object is placed at different distances	Identify the position of the image formed by a lens using pre-existing rays on a diagram	**Construct ray diagrams showing the formation of images by a convex lens when the object is placed at different distances**	Fully describe the properties of an image (real, virtual, magnified, diminished, upright, and inverted) based on a ray diagram for a convex lens
Construct ray diagrams showing the formation of images by a concave lens	State that concave lenses only produce virtual images	**Construct ray diagrams showing the formation of images by a concave lens**	Fully describe the properties of an image based on a ray diagram for a concave lens
Use ray diagrams to describe the similarities and differences between concave and convex lenses	State some uses of lenses	**Use ray diagrams to describe the similarities and differences between concave and convex lenses**	Use ray diagrams to determine the magnification of the lens

Tier 2 vocabulary	Tier 3 vocabulary
light	converge, convex, focal length, focus, image, lens, magnify, ray diagram, *virtual*

Lesson plan

Reactivate knowledge

1 What happens to parallel rays of light as they enter a concave lens?

2 Give one use of a convex lens.

3 How do you determine the magnification of a lens?

Answers: 1 The light diverges/spreads out as if it came from a focus in front of the lens *2* Any from: camera, magnifying glass, projector *3* Image height ÷ object height

Common misconception

Students may think that when a lens is moved, an image will become bigger or smaller but will always remain sharp. Explain to students that lenses work by changing the direction of light as it passes through it and focusing the light (bringing all rays together) at a specific point. If the lens is moved, the point of focus will also change, so the image formed may be blurred.

Trigger interest

Magnified mysteries (5 mins)

Show a selection of highly magnified images of mystery objects for students to identify, for example, a fly's eye, human hairs, Velcro hooks, a rough surface, and hairs on a gecko's feet.

Use this to recap the concept of magnification, including a short calculation.

Exposition

Real image from a convex lens (30 mins)

Construct a ray diagram from scratch on a large white-board or under a visualiser, as you explain each part. Students should follow along, using large (e.g., A3 sized) paper and 30 cm rulers. This is time-consuming, but is critical to understanding image formation.

Questions can be asked along the way, to ensure students recap the ideas of converging rays and focal lengths from P14.4.

Then, demonstrate the adjustment of a lens to form a clear image on the screen described in the practical demonstration. Compare this with the movement of a lens on a real camera (a video clip of this is useful).

Once the demonstration is complete, return to the board or visualiser and repeat the process with the students to show the construction of a ray diagram for a virtual image from a lens. Discuss the difference between a real image and a virtual image again.

Ensure students practise additional construction of appropriate diagrams, or identification of correct rays and images.

Main activity

Real image from a convex lens (10 mins)

Ensure the screen is set at a fixed position on the ruler or optical bench. Use a converging lens (focal length approximately 10 cm) to demonstrate the formation of a real image and the repositioning required to form a clear image.

Review and practice

Camera developments (10 mins)

Discuss the advantages and disadvantages of film-based technology in comparison to digital photography, such as that used in modern smartphones. A short discussion of the evolution of photography works well.

Support this discussion by showing images of cameras throughout history and emphasise how ubiquitous lenses are in the modern era. If there is time, extend the discussion by asking students for their opinions on the social and ethical implications of this fact, including reference to the reduction in size and drastic improvement in functionality.

P14 Light

Introduction to checkpoint intervention

The checkpoint intervention provides suggestions for a lesson to follow up the *P14 Light* checkpoint assessment. Depending on the outcomes of the assessment, these suggestions can be used to move a class towards secure, to target specific learning outcomes, or to consolidate knowledge for students achieving secure outcomes. Use students' outcomes from the checkpoint assessment to address general misconceptions from the content covered. Be prepared to re-cover content as required; three key concepts from the chapter are covered in more detail below.

Checkpoint secure learning outcomes

- Draw diagrams to show reflection from smooth and rough surfaces.
- Investigate the refraction of light through a glass or Perspex block.
- Determine the apparent colour of a coloured surface when illuminated by different colours of light.
- Describe the properties of an image formed in a mirror.
- Construct ray diagrams showing the formation of images by a convex lens.
- Calculate the magnification of a lens based on object and image size.

Student reflection

Encourage students to reflect on whether there were any checkpoint questions they found difficult or straightforward as well as their preparation for the checkpoint assessment, including how they revised, what they revised, the time spent on revision, and what areas they could improve upon. For more strategies, see the Metacognition spread.

Intervention activity: Investigate the refraction of light through a glass or Perspex block.

PhET offer a great virtual toolkit to explore refraction effects through glass. Students can build their confidence using this simulation, or another, to explore what happens to light rays as they enter glass or Perspex. Note that the simulation does not have arrows – students will need to add arrows to show the direction of their rays for full credit. Compare these simulations with real effects, using ray boxes and glass or Perspex.

Allow and praise loose descriptions (e.g., *the light ray bends* or *changes direction*) before developing these responses and carrying out more quantitative analyses. Reinforce the definition of refraction as the change of speed of light as it moves from one material to another, which often results in a change of direction.

Once students are able to recognise refraction effects, they are ready to quantify them using angles. Frame this as a relationship between the incoming angle (angle of incidence) and the outgoing angle (the angle of refraction), focusing on the air–glass boundary.

Refer to a diagram of this, carefully labelling the normal line, and ask students which equipment might be helpful to measure angles. Model how these angles can be measured directly, using the ray box and glass, while students follow along using their own equipment. Ensure pairs of students swap roles to measure angles of incidence and refraction separately; collect results together in a large results table, leading students to conclude that the angle of incidence is always greater than the angle of refraction, when light goes from air to glass.

Intervention activity: Describe the properties of an image formed in a mirror.

Show students their reflection in a plane mirror, and ask them to describe what they notice, given some prompts: size of the image compared to their real size, the distance to the mirror and to the image behind the mirror, the orientation of the image (upright, or upside down). Discuss their observations before they write them down.

Once students are able to describe the properties of their image formed in a mirror, give them a ray diagram that shows how a virtual image is formed in a mirror. Emphasise the use of dotted lines to illustrate the fact that light is not really originating from the image, which leads to the definition of virtual images.

Using the same diagram, students should then practise sketching the rays from another object, reflecting at the mirror and entering their observing eye. Make sure they can trace these rays to an image behind the plane mirror.

Intervention activity: Calculate the magnification of a lens based on object and image size.

The fact that light (on ray diagrams) starts from the object and finishes at an image (usually on a screen) is not always obvious to students. Begin this intervention activity with a selection of ray diagrams showing images being formed. Students should label the objects and the images and explain how they know which is which. Emphasise the word 'focus' at the point where images are formed.

Use one of the diagrams from the starter that shows a magnified image, and model how the equation for magnification is used. Students should understand that the value is a ratio, and that numbers greater than one (>1) mean the image is larger than the object, which corresponds to a magnifying glass. Mathematical reasoning (that the numerator is larger than the denominator) can help to reinforce this property. Give students a couple of other simple calculations, ensuring they use a ruler to measure image and object heights directly.

P15 Electromagnetism

P15.1 Magnetic fields

AQA GCSE Physics specification points

4.7.1.1 Poles of a magnet

4.7.1.2 Magnetic fields

***Smart Curriculum* Narrative**

Electricity and magnetism

Theme/subtheme

Electricity and magnetism/Magnetism

Disciplinary knowledge

- WS 2.2

Practical

Magnets and magnetic fields

	Learning outcomes		
Learning objectives	**Developing Aiming for 3**	**Secure Aiming for 5**	**Extending Aiming for 8**
Describe the difference between permanent magnets and induced magnets	Describe the forces between like and unlike poles of permanent magnets	**Describe the difference between permanent magnets and induced magnets**	Explain in detail how magnetism can be induced in some materials
Explain how the behaviour of a compass is evidence for the magnetic core of the Earth	Compare the Earth's magnetic field to that of a bar magnet	**Explain how the behaviour of a compass is evidence for the magnetic core of the Earth**	Explain why it would be difficult to use a compass at the magnetic poles of the Earth
Draw the magnetic field around a bar magnet including the direction of the magnetic field lines	Describe how to plot the magnetic field of a magnet using a compass	**Draw the magnetic field around a bar magnet including the direction of the magnetic field lines**	Describe the regions in a magnetic field where magnetic forces are greatest using the idea of field lines

Tier 2 vocabulary	Tier 3 vocabulary
alloy, force, magnet, magnitude, metal, permanent, radiation, steel	attract, bar magnet, contact force, field lines, force field, magnetic field, non-contact force, north pole, plotting compasses, repel, south pole

Lesson plan

Reactivate knowledge

1 Is magnetism a contact or non-contact force?

2 What happens when you bring the S-pole of a magnet near to a S-pole of a second magnet?

3 What is a magnetic field?

Common misconception

Students may think that all metals are attracted to a magnet. Only a few metals including iron, nickel, cobalt, and their alloys are attracted to magnets. For example, copper, silver, gold, and aluminium are all metals, but they are not magnetic. In fact, most metals are not naturally magnetic.

Answers: 1 Non-contact *2* They repel *3* A region where a magnet or magnetic material feels a force

Trigger interest

Force fields (10 mins)

Interactive activity where students categorise – into contact and non-contact forces – a list of the various forces they have studied.

Recap key ideas about the different forces with individual students as appropriate, extending discussions to explore with students what factors affect the magnitude of the various forces.

Exposition

Magnets and compasses (15 mins)

Much of this lesson can be supplemented by live demonstrations using compasses, bar magnets, and iron filings, if available. Visualisers work well so that key ideas can be observed alongside verbal explanations, but annotated images (e.g., of iron filings aligned to the magnetic field around a bar magnet) also work well.

Using two bar magnets, demonstrate the general rule of magnetism – student volunteers can try to push two like poles together to reinforce the effects of magnetic repulsion. Ensure students appreciate that 'repel' means to push apart, while 'attract' means to pull together, which may not be obvious.

Supplementing this demonstration with a diagram, showing the direction of the repulsive force pairs on two north poles and two south poles, and the attractive force pair on a north and south pole, is useful. Emphasise that iron and steel objects can be magnetised (i.e., turned into magnets), but permanent magnets are made of steel because magnetised steel does not lose its magnetism easily.

A magnetic field should be explained as an invisible region around a magnet in which other magnetic materials experience a force. Students should observe the effect of a magnetic field by observing how it causes iron filings to align along the direction of field lines – responding clearly to this 'magnetic force field'. Ensure this effect is clearly visible and reinforce the pattern by comparing to a labelled image or diagram.

Main activity

Magnets and magnetic fields (20 mins)

In this practical, pairs of students investigate the properties of magnets. Circulate the room and discuss the interaction of the bar magnet (or the needle in a compass) with the Earth's magnetic field. Students should also get experience plotting the field around the bar magnet using plotting compasses – two or three field lines is sufficient if time is short.

Demonstrate induced magnetism by lifting a paperclip with a strong bar magnet. Students should first be convinced that the paperclip itself is not magnetic but becomes magnetised in the presence of another magnet.

Extend discussions by asking students to guess where the strength of the field might be strongest – leading them to recognise it is where field lines are closer together, i.e., near the poles of the magnet.

Review and practice

Earth is a protective magnet (10 mins)

Show students an image of the magnetic field around the Earth, such as Figure 4 in the Student Book. Students should recognise the shape of the field as being similar to that around a bar magnet. Discuss how the Earth's magnetic field and its magnetic poles, which tend to gradually move around, provide evidence for a metallic material in a molten core. As an extension, students should consider the almost totally vertical field lines found at the Earth's poles, and use this to explain why it would be difficult to use a compass at this location.

Finally, use a video clip or animation of the solar wind, discussing how the magnetic field around the Earth protects the planet from potentially harmful radiation and energetic particles.

P15.2 Magnetic fields of electric currents

AQA GCSE Physics specification points

4.7.2.1 Electromagnetism

Smart Curriculum Narrative

Electricity and magnetism

Theme/subtheme

Electricity and magnetism/Magnetism

Disciplinary knowledge

- WS 2.2
- MS 4a

Practical

Fields around a current-carrying wire

	Learning outcomes		
Learning objectives	**Developing Aiming for 3**	**Secure Aiming for 5**	**Extending Aiming for 8**
Describe how the current in the wire and the distance from the wire affect the strength of the magnetic field	Describe an experiment to show the magnetic field around a current-carrying wire	**Describe how the current in the wire and the distance from the wire affect the strength of the magnetic field**	Explain the factors affecting the strength of the magnetic field around a wire
Draw the magnetic field pattern (including the direction) produced by a straight wire carrying a current and for a solenoid	State that the magnetic field produced by a current-carrying wire is circular	**Draw the magnetic field pattern (including the direction) produced by a straight wire carrying a current and for a solenoid**	Explain how the magnetic field pattern of a solenoid shows the magnetic field inside the solenoid is strong and uniform
Explain why coiling a wire to form a solenoid increases the strength of the magnetic field	State the factors that increase the strength of the magnetic field around a solenoid	**Explain why coiling a wire to form a solenoid increases the strength of the magnetic field**	Explain why iron and not steel is used in the core of a solenoid to increase the strength of the magnetic field

Tier 2 vocabulary	Tier 3 vocabulary
current, electricity, magnet, wire	coil, corkscrew rule, electromagnet, loop, magnetic field, solenoid

Lesson plan

Reactivate knowledge

1 How can you make a magnetic field visible?

2 What is an induced magnet?

3 How do you make an electromagnet?

Answers: 1 Using plotting compasses or iron filings *2* A magnetic material that is magnetised by a magnetic field *3* Wrap a current-carrying coil or wire round an iron core

Common misconception

Students may think that big magnets are stronger than small magnets. Tell students that big magnets may not necessarily be stronger than smaller magnets – this will depend on the material they are made from. For example, a small neodymium magnet can be much stronger than a large iron magnet.

Trigger interest

Electricity and magnetism recap (10 mins)

Students complete the interactive with a set of electricity questions to establish and recap prior knowledge.

Exposition

Wires, currents, and fields (15 mins)

Practical demonstrations of the effects described in this lesson work best, whether being observed by the whole group (e.g., by eye or by use of a visualiser) or explored independently by the students themselves, ideally in pairs, if the equipment is available.

The key idea is that any current-carrying wire has a magnetic field around it. Another way to express this is that a current-carrying wire acts like a small magnet. Ensure students appreciate that the magnetic field around such a wire is relatively weak and can be made stronger by turning the wire into a solenoid. Demonstrate the deflection of a plotting compass as it is moved around a powered solenoid. This can lead nicely into a short explanation of electromagnets.

Extend these practical demonstrations by discussing the historical significance of these findings – at the time of discovery (especially the induction effects to come in P15.5) they were nothing short of magical, and laid the groundwork for eventual revolutions in technology and engineering. Students can think about some of the practical applications of electromagnetism as they explore the various effects in the practical.

Main activity

Fields around a current-carrying wire (20 mins)

In this practical, students plot the circular magnetic field around a current-carrying wire. Ensure that pairs of students take turns plotting the direction of the field as the current is reversed, and recognise the effect on the direction of the magnetic field. In addition, ensure all students can correctly apply the corkscrew rule to determine the direction of a magnetic field around a current-carrying wire.

Review and practice

More power (10 mins)

Ask students to outline a test to explore what affects the strength of a magnetic field around a solenoid. List the factors (the current, the number of loops in the coil, and the type of material), from which they should choose one factor to investigate. Students can choose if their plan explores a qualitative or quantitative relationship and share their approaches with a partner.

P15.3 Electromagnets in devices

AQA GCSE Physics specification points

4.7.2.1 Electromagnetism

Smart Curriculum Narrative

Electricity and magnetism

Theme/subtheme

Electricity and magnetism/Magnetism

Disciplinary knowledge

- WS 1.4
- MS 4a

Practical

Investigating electromagnets

	Learning outcomes		
Learning objectives	**Developing Aiming for 3**	**Secure Aiming for 5**	**Extending Aiming for 8**
Study a diagram of an electromagnetic device and explain how it works	Identify the electromagnet in a diagram of an electromagnetic device	**Study a diagram of an electromagnetic device and explain how it works**	Study a diagram of an electromagnetic device and suggest improvements to design
Evaluate the quality of data collected for electromagnet investigation in terms of the precision	Identify anomalies in data collected for electromagnet investigation	**Evaluate the quality of data collected for electromagnet investigation in terms of the precision**	Evaluate the quality of data and suggest improvements to the method

Tier 2 vocabulary	Tier 3 vocabulary
force, magnet, metal, wire	electromagnet, *field*, loop, magnetic field, magnetisation, solenoid

Lesson plan

Reactivate knowledge

1 How do you increase the strength of an electromagnet?

2 What is needed to make an electromagnet?

3 Name three magnetic materials.

Common misconception

Students may think that all metals are attracted to a magnet. Only a few metals including iron, nickel, cobalt and their alloys are attracted to magnets. For example, copper, silver, gold, and aluminium are all metals, but they are not magnetic. In fact, most metals are not naturally magnetic.

Answers: 1 Increase current, increase number of coils *2* Current-carrying coil of wire and iron core *3* Any three from: iron, steel, cobalt, or nickel

Trigger interest

On and off again (5 mins)

Give a short demonstration of an electromagnet lifting a long chain of (steel) paperclips. Each paperclip can be added one-by-one for added effect. Students should discuss and explain what they observe in terms of magnetisation and forces (i.e., the electromagnet magnetises the steel in the paperclips, and therefore unlike poles attract each end together).

Extend this demonstration using a short video clip that shows a similar effect, for example, of electromagnets used to sort metal objects in a scrap yard. Be aware of misconceptions about metals and magnetism – remind students that only iron, steel, nickel, and cobalt (plus their alloys) are magnetic.

Exposition

Solenoids and electromagnets (10 mins)

Go over the key ideas about the magnetic fields around solenoids. Discuss the current through each loop of wire, the shape of the magnetic field, and the factors that affect its strength. Use mini-whiteboards to assess students' understanding.

Next show students two side by side images of an application of electromagnets (e.g., in an electric bell vs. in a relay switch). Split the class into two groups, assigning one device to each group. Ask students to think, discuss, and share ideas about how the principles of electromagnetism allow their device to work. The key ideas should centre on the flowing currrent, an associated magnetic field, magnetisation, and magnetic attraction.

Main activity

Investigating electromagnets (30 mins)

Students investigate electromagnets and the practical applicability of the fields produced, testing strength and the factors that affect it. Ensure that roles are rotated (e.g., circuit builders, describers, explainers). Students could investigate one or two of the applications, then collate their results in groups, or present their results in turn to their classmates.

Review and practice

Marvellous magnets (10 mins)

Show students short video clips about how electromagnetic principles have been used in transport systems (e.g., mag-lev trains), in cutting-edge scientific research (e.g., the *Large Hadron Collider*), or in medicine (e.g., MRI scanners). Discuss the advantages and challenges associated with one or more of these systems.

Ⓗ P15.4 The motor effect

AQA GCSE Physics specification points

Ⓗ 4.7.2.2 Fleming's left-hand rule

Ⓗ 4.7.2.3 Electric motors

Smart Curriculum Narrative

Electricity and magnetism

Theme/subtheme

Electricity and magnetism/Magnetism

Disciplinary knowledge

- MS 3b, 3c

Practical

The motor effect

Learning objectives	Learning outcomes: Developing Aiming for 3	Learning outcomes: Secure Aiming for 5	Learning outcomes: Extending Aiming for 8
Apply Fleming's left-hand rule to determine the direction of the force acting on a conductor	State that the motor effect is when a current-carrying conductor is placed in a magnetic field and they both feel a force	**Apply Fleming's left-hand rule to determine the direction of the force acting on a conductor**	Explain why there is a force acting on a conductor in a magnetic field
Calculate the force acting on a conductor when it is placed in a magnetic field	State the factors that affect the size of the force acting on a conductor in a magnetic field	**Calculate the force acting on a conductor when it is placed in a magnetic field**	Perform calculations involving rearrangements of the equation $F = BIl$
Explain why the coil of a motor rotates	State that a motor consists of a coil of wire carrying a current which rotates in a magnetic field	**Explain why the coil of a motor rotates**	Explain the factors that effect the speed of rotation of the coil

Tier 2 vocabulary	Tier 3 vocabulary
current, force, magnet, proportional, wire	coil, conductor, electromagnetism, *field*, Fleming's left-hand rule, magnetic field, motor effect

Lesson plan

Reactivate knowledge

1 What is the direction of a magnetic field?

2 What happens to the magnetic field around a wire when you change the current direction?

3 How do you increase the strength of a magnetic field around a wire?

Common misconception

Students may think that the magnetic force on a current-carrying coil of wire is a small effect. It can actually be made very large because it depends on the strength of the magnetic field the coil is in (B), the total length of wire in the field (l), and the size of current flowing in the wire (I). It is therefore expressed by the equation $F = BIl$.

Answers: 1 From N-pole to S-pole /direction of force a N-pole would feel if placed at that point *2* Magnetic field changes direction *3* Increase the current/coil the wire

Trigger interest

Magical motor effect (5 mins)

Demonstrate the effect of a 'kicking wire' (i.e., that there is a force exerted on a current-carrying conductor in a magnetic field), and the factors that affect the size of this effect. Students should appreciate the natural wonder of this demonstration – a direct link between electromagnetism and motion – and how it played a vital role in kicking off the Industrial Revolution. Describe how the outcomes of these experimental discoveries, primarily by Michael Faraday, played a vital role in the second industrial revolution, sometimes known as the technological revolution.

Exposition

Fleming's Left-Hand Rule (15 mins)

Show how Fleming's left-hand rule can be used to determine the direction of the force acting on a current-carrying conductor in a magnetic field. Provide plenty of practise for students to determine the direction of the force (the direction of motion), using simple 2D diagrams of wires in magnetic fields, before moving on. It can be helpful to introduce language (and symbols) for motion 'into the page' and 'out of the page'.

Extend these examples by asking students to determine the direction of the magnetic field (given force and current) and/or the direction of the current (given force and magnetic field).

Introduce the equation $F = BIl$ as a means to calculate the size of the force. Define each of the symbols, and provide a couple of simple examples, explicitly showing how to substitute the correct values. Provide calculation frames where needed.

Link this equation to the demonstration of the kicking wire – the force experienced by the wire is proportional to the strength of the external field, the length of wire in this field, and the amount of current through it.

Main activity

The motor effect (25 mins)

Demonstrate the motor effect using the practical provided. If time permits, students can construct small model motors from standard kits. To save time, and for additional support where needed, some students can be provided with partly completed motors to finish, for example, having pre-wound coils. Wherever possible, all students should observe the operation of a fully completed motor.

Where needed, students can watch video clips that animate the motor effect to show how the rotation is produced. It is important they can see that opposite sides of the coil of wire in the magnetic field experience forces in opposite directions, which causes rotation. Students should also observe, using their models or by simulation, that increasing the current supplied/turns of wire in the coil increases the rotation speed, because the size of the force on the wires increases.

Review and practice

The motor effect (10 mins)

Students use the interactive to complete a paragraph to describe the motor effect. They then carry out a few calculations to embed the equation.

Ⓗ P15.5 The generator effect

AQA GCSE Physics specification points

Ⓗ 4.7.3.1 Induced potential

Smart Curriculum Narrative

Electricity and magnetism

Theme/subtheme

Electricity and magnetism/Electromagnetic induction and the generator effect

Disciplinary knowledge

- WS 1.4
- MS 4a

Practical

Investigating a simple generator

	Learning outcomes		
Learning objectives	**Developing Aiming for 3**	**Secure Aiming for 5**	**Extending Aiming for 8**
Describe electromagnetic induction in a wire	State that the generator effect is when there is relative movement between a conductor and a magnetic field inducing a potential difference in the conductor	**Describe electromagnetic induction in a wire**	Explain why relative movement of a wire through a magnetic field or a changing magnetic field is required to cause induction
Describe how to increase the size of an induced potential difference	State the factors that affect the size of an induced potential difference	**Describe how to increase the size of an induced potential difference**	Explain the factors affecting the size of the induced potential difference
Explain how the direction of the induced current relates to the magnetic field	State that the direction of the induced current always opposes the change that caused it	**Explain how the direction of the induced current relates to the magnetic field**	Determine the change in direction of the induced current from information provided

Tier 2 vocabulary	Tier 3 vocabulary
circuit, *conductor*, *current*, force, magnet, parallel, perpendicular, *work*	conservation, electromagnetic, galvanometer, generator, induction, magnetic field

Lesson plan

Reactivate knowledge

1 Wind turbines contain electricity generators and can only generate electricity when it is windy. What is the input energy store of a generator?

2 What is the motor effect?

3 Why is a compass affected by electrical wires?

Common misconception

Students may think that when a conductor moves through a magnetic field, a current is induced in the wire. Respond that firstly, relative motion between a conductor and a magnetic field induces a potential difference, but not necessarily a current. A current is induced (because of an induced potential difference) only if the conductor is part of a complete circuit. Secondly, if the conductor moves parallel to magnetic field lines, then no potential difference is induced. It must move perpendicularly to the magnetic field lines to induce the maximum potential difference.

Answers: 1 Kinetic *2* When a current-carrying wire feels a force in a magnetic field *3* There are magentic fields around the current in the wires

Trigger interest

Generate ideas (5 mins)

A large model generator can be used to help students grasp the key idea that relative motion between magnets and conductors can induce a potential difference (or current, if using a wind-up torch, for instance). Show students a video clip of a large generator in operation in a power station. Students should list the key components they see and explain why the generator rotates.

Encourage students to make connections, or show explicit links to P3.2, which touched on the operation of wind turbines to generate electricity. Students could also suggest what would happen if all generators in the UK suddenly stopped working – would sort of impact would this have on our everyday lives?

Exposition

Exploring induced potential (15 mins)

Wherever possible, give exposition of this theory in small groups. Induction is best observed and explored experimentally, so give as much time as possible for the students to have hands-on experience with the practical, and answering structured questions about what they observe. An introductory demonstration of electromagnetic induction, using a large galvanometer (a very sensitive ammeter, e.g., with a moving dial visible to students, or shown under a visualiser, where available). Introduce the term 'induction' and ask students to predict what types of things could increase the size of the induced potential difference.

As you circulate, the key learning concept to assess is that relative motion between a conductor and a magnet will create ('induce') a potential difference – across the ends of the conductor. Be sure to use questions that explore the factors affecting the size and direction of the induced potential difference, and whether the potential difference results in an induced current (i.e., if the conductor is part of a complete circuit).

Main activity

Investigating a simple generator (25 mins)

The students test out factors as described in the practical, noting changes in the induced current.

Demonstrate, or allow early finishers to test induction in a solenoid as described in the practical.

They should identify all possible combinations (i.e., polarities at the ends of the solenoid), and appreciate that there must always be relative motion between the wire and the magnet to produce the current.

Review and practice

Conservation (10 mins)

Students use the interactive to explain how they know that energy is conserved in all of these generator tests by using the idea of work done by a force, and work done by a current.

Extend this activity by asking students the factors they would need to know to calculate the work done in each case.

(H) P15.6 The alternating-current generator

AQA GCSE Physics specification points

(H) 4.7.2.4 Loudspeakers

(H) 4.7.3.2 Uses of the generator effect

(H) 4.7.3.3 Microphones

Smart Curriculum Narrative

Electricity and magnetism

Theme/subtheme

Electricity and magnetism/Electromagnetic induction and the generator effect

Disciplinary knowledge

- WS 1.4
- MS 4a

Activity

Alternators and generators

	Learning outcomes		
Learning objectives	**Developing Aiming for 3**	**Secure Aiming for 5**	**Extending Aiming for 8**
Explain how an alternator generates a.c.	State that an alternator generates a.c. and a dynamo generates d.c.	**Explain how an alternator generates a.c.**	Explain how a dynamo generates d.c.
Draw potential difference against time graphs for the output of a generator, identifying peak potential difference and period of the cycle	State that increasing the speed of rotation, the number of coils, the area of the coils, and the strength of the magnetic field all increase the peak potential difference	**Draw potential difference against time graphs for the output of a generator, identifying peak potential difference and period of the cycle**	Determine the position of the coil of an alternator from a potential difference time graph of its output
Describe how microphones use the generator effect and loudspeakers use the motor effect	State loudspeakers use the motor effect and microphones use the generator effect	**Describe how microphones use the generator effect and loudspeakers use the motor effect**	Explain how a moving coil microphone and moving coil loudspeaker work

Tier 2 vocabulary	Tier 3 vocabulary
coil, *current*, energy, negative, positive, *wave*, wire	alternator, antennae, *diaphragm*, direct, electromagnetic, generator effect, loudspeaker, magnetic field, motor, oscillate, oscilloscope, *trace*

Lesson plan

Reactivate knowledge

1 Describe one way to induce a potential difference.
2 How do you increase the induced potential difference?
3 What is the difference between a.c. and d.c?

Common misconception

Students may think that an oscillating waveform indicates an alternating p.d. or a.c. A source of d.c. might vary its output between 1 A and 2 A. However, a source of a.c. varies its output positively and negatively, for example, between 1 A and –1 A. This corresponds to oscillations (changing direction) for the p.d. or current. Therefore, a varying *direct p.d. or current* can look similar to an *alternating p.d. or current* (e.g., on an oscilloscope trace) but they are not equivalent.

Answers: 1 Move a magnet near a coil/move a coil near a magnet *2* Move the magnet faster/have more coils/increase the strength of the magnetic field *3* Alternating current changes direction continuously, d.c. has the same direction

Trigger interest

d.c. or a.c.? (5 mins)

Show students two sketched graphs of a varying d.c. (always positive) and an a.c. (alternates positively and negatively). Though these look similar, they are not equivalent. Ask students to discuss, then vote which is which.

This will highlight a common misconception which will help in the Exposition. Students should be able to distinguish a varying d.c. from an a.c.

Exposition

Alternators vs. Generators (15 mins)

As with topic P15.5, create plenty of time for practical demonstration and/or hands-on exploration for this lesson as possible. Start with a recap of P15.5, showing a demonstration of the generator effect and the fact that moving a wire or coil 'to-and-fro' inside a magnetic field, will generate an alternating p.d.. This is the central learning idea for this lesson.

Show students an a.c. generator and compare its construction to a motor. They should note similarities (coil of wire, the magnetic field), and the main difference – the absence of a power source (because a p.d. is going to be *generated*). Emphasise that the generator coil must be *made* to rotate, linking to P15.5.

Show the a.c. output of a generator. Ask students why the current changes direction (alternatively, that the alternating p.d. is both positive and negative). Link this to the brushes and the changes of connection. A suitable animation or video works well here.

Main activity

Alternators and generators (25 mins)

Students complete the worksheet sketching the oscilloscope trace and answering questions related to the generation of a.c. or d.c. current and the construction of the alternator/generator. The demonstration can be repeated for small groups, or alternatively, students can watch various animations as research to help answer the questions.

Review and practice

Microphones and loudspeakers (10 mins)

Show a deconstructed coil microphone, if available, or an annotated image. Link its operation to the generator effect, which can be summarised as motion (i.e., the vibrating diaphragm) converted to an alternating electrical signal (i.e., via the generator effect). Extend this discussion by linking this idea to electromagnetic waves causing the oscillation of electrons in car radio antennae. Ask students to consider the similarities in how these processes transfer energy.

H P15.7 Transformers

AQA GCSE Physics specification points

4.7.3.4 Transformers

Smart Curriculum Narrative

Electricity and magnetism

Theme/subtheme

Electricity and magnetism/Transformers

Disciplinary knowledge

- WS 1.4

Practical

Making a model transformer

	Learning outcomes		
Learning objectives	**Developing Aiming for 3**	**Secure Aiming for 5**	**Extending Aiming for 8**
Describe the structure of a transformer	State that transformers consist of two coils wound on an iron core	**Describe the structure of a transformer**	Justify the choice of materials used to construct a transformer
Describe the operation of a transformer in simple terms	State that transformers need an a.c. supply	**Describe the operation of a transformer in simple terms**	Explain how an a.c. current in the primary coil induces an a.c. current in the secondary
Describe the difference between a step-up and a step-down transformer in terms of the number of turns on the primary and secondary coils	Describe how step-up transformers are used to increase potential difference and step-down transformers to decrease the potential difference	**Describe the difference between a step-up and a step-down transformer in terms of the number of turns on the primary and secondary coils**	Explain why having more coils on the secondary than the primary steps up the potential difference

Tier 2 vocabulary	Tier 3 vocabulary
coil, *current*, electric, proportional, ratio, voltage, wire	alternating current, input, induction, magnetic field, output, primary coil, secondary coil, step-down, step-up, transformer

Lesson plan

Reactivate knowledge

1 Suggest a place where transformers are used.

2 Describe the magnetic field around an alternating current.

3 What does a step-up transformer do?

Common misconception

Students may think transformers operate using any form of electric current. Respond that they typically require an a.c.. This is because the secondary coil of wire requires a changing magnetic field, otherwise, no p.d. can be induced in it. The only way to set up a changing magnetic field is to ensure there is an a.c. in the primary coil of wire.

Answers: 1 Any sensible suggestion, for example, power stations/chargers etc. *2* The magnetic field is alternating/changing size and direction *3* Increases the output p.d.

Trigger interest

Supplies supplies! (5 mins)

Show students some example power supplies used for phones, laptops, and so on. They should note that although they operate from a 230V a.c. supply, they have a range of outputs. Ask students for their ideas about what mechanisms inside these devices might 'transform' these potential differences.

Exposition

The transformer (10 mins)

As with P15.5 and P15.6, spend the majority of this lesson using the practical to explore the key learning concepts. This lesson brings all of the central electromagnetism ideas together – it will be beneficial to recap, or assess these in turn with small groups as needed.

They are:

- iron can be magnetised (P15.1)
- an (alternating) current creates its own (alternating) magnetic field (P15.2)
- a wire in an alternating magnetic field induces an alternating p.d., because there is relative change ('motion') between them (P15.5)
- the size of the alternating p.d. is proportional to the length of wire (i.e., the number of coils) (P15.5).

A large, labelled diagram of a transformer is an effective introduction to the practical activity, introducing the vocabulary of 'primary' and 'secondary' coils, and the 'input' and 'output' p.d. (or current) across (or through) these coils.

Main activity

Making a model transformer (35 mins)

Students construct their own transformers, describe how they operate and, if time permits, explore the relationship between the coil and the voltage ratio.

Use the opportunity during their construction to circulate the room and assess each of the key ideas outlined above. Students should appreciate why the transformer will not work with direct current and explain that *changing* magnetic fields are required for induction in the secondary coil.

Review and practice

Revisiting the National Grid (5 mins)

Remind students of the use of transformers in the National Grid, from P5.1. Show a simplified diagram of the National Grid, highlighting the step-up and step-down transformers. Discuss how and why these change the voltage, emphasising the constant power transmission, so high transmitting voltage means a low current, and vice versa.

Students should appreciate that a low current means less energy is wasted to the thermal energy stores of the surroundings, which is useful as these transmission cables typically cover large distances. Refer to the pylons that are used to elevate the unsafe, high-voltage wires above ground. Referring to step-down transformers, students should link back to the activity in the Trigger interest.

H P15.8 Transformers in action

AQA GCSE Physics specification points

4.7.3.4 Transformers

Smart Curriculum Narrative

Electricity and magnetism

Theme/subtheme

Electricity and magnetism/Transformers

Disciplinary knowledge

- MS 1c, 3b, 3c

Activity

Transformers and efficiency

	Learning outcomes		
Learning objectives	**Developing Aiming for 3**	**Secure Aiming for 5**	**Extending Aiming for 8**
Use the transformer equation to calculate input or output voltages for a transformer	State that the ratio of the primary and secondary potential difference depends on the ratio of the number of turns on the coils	**Use the transformer equation to calculate input or output voltages for a transformer**	Apply the transformer equation in a wide variety of situations
Use the output power to determine the primary current	State that for a 100% efficient transformer, the input power equals the power output	**Use the output power to determine the primary current**	Solve problems using both power and transformer equations
Apply transformer equations to explain the advantage of high voltage transmission	State that the heating effect of a current in a wire is proportional to the square of the current	**Apply transformer equations to explain the advantage of high voltage transmission**	Solve complex problems involving high voltage transmission

Tier 2 vocabulary	Tier 3 vocabulary
current, dissipation, efficiency, electric, energy, *power*, wire	alternating current, input, magnetic field, motor effect, national grid, output, potential difference, primary coil, secondary coil, step-down, step-up, transformer

Lesson plan

Reactivate knowledge

1 What is the basic structure of a transformer?

2 How do the number of coils on the primary and secondary compare for a step-down transformer?

3 How do you calculate electrical power?

Common misconception

Students may think step-up transformers increase the voltage and the current. Respond that they increase the voltage at which electrical power is transmitted. Since the voltage increases for a given amount of power transmission, the current that delivers this power can reduce, because power $P = V$ (voltage) × I (current). Similarly, step-down transformers decrease the voltage for power transmission, so the current increases.

Answers: 1 Primary and secondary coil and iron core *2* More coils on the primary than the secondary *3* Potential difference × current

Trigger interest

Higher or lower? (5 mins)

Show students a selection of transformer arrangements (various step-up and step-down transformers). Using this opportunity to recap the key vocabulary from P15.7 (i.e., primary and secondary coils, and input vs. output p.d.), ask students to vote, using mini-whiteboards, whether the output p.d. will be higher or lower than the input p.d.

Exposition

The transformer equation (10 mins)

Discuss the factors that affect transformers (number of turns and potential differences), leading to the transformer equation. Ensure students are comfortable recognising the subscripts (the subscript p stands for the primary coil, and the subscript s stands for the secondary coil).

Main activity

Transformers and efficiency (35 mins)

After an example using the transformer equation, showing clear substitution of relevant values, the students should have some time to practise using this equation. Ensure that they try a calculation to find each variable at least once – support this activity using a calculation frame (especially when rearranging), wherever needed.

Discuss the potential sources of energy dissipation in the transformer. State that most transformers are nearly 100% efficient, before asking students to recap what 'efficiency' means.

Present the equation for transformer power, emphasising how the power generated in the secondary coil is equal to the power generated in the primary coil. Lead students to appreciate that this equation assumes 100% efficiency. As before, show an example using this equation, explaining the reasons for high-voltage and low-current transmission in the National Grid.

Since $P = IV$, power can be transmitted at high voltage with low current, to reduce the heating effect in transmission wires. This heating effect, due to a current, is large because $P = I^2R$. Therefore, small currents are desirable in power transmission.

Students can use the calculation worksheet: *The motor effect and transformers* to explore more mathematical calculations in greater depth.

Review and practice

Transformer matching (5 mins)

Interactive activity where students match up transformers with changes in p.d.

Support this activity by providing a set of possible ratios to choose from for each transformer.

P15 Electromagnetism

Introduction to checkpoint intervention

The checkpoint intervention provides suggestions for a lesson to follow up the *P15 Electromagnetism* checkpoint assessment. Depending on the outcomes of the assessment, these suggestions can be used to move a class towards secure, to target specific learning outcomes, or to consolidate knowledge for students achieving secure outcomes. Use students' outcomes from the checkpoint assessment to address general misconceptions from the content covered. Be prepared to re-cover content as required; three key concepts from the chapter are covered in more detail below.

Checkpoint secure learning outcomes

- Draw the magnetic field around a bar magnet including the direction of the magnetic field lines.
- Study a diagram of an electromagnetic device and explain how it works.
- Apply Fleming's left-hand rule to determine the direction of the force acting on a conductor.
- Describe how microphones use the generator effect and loudspeakers use the motor effect.
- Draw potential difference against time graphs for the output of a generator, identifying peak potential difference and period of the cycle.
- Describe how the primary current compares to the secondary current in a step-up transformer.

Student reflection

Encourage students to reflect on whether there were any checkpoint questions they found difficult or straightforward as well as their preparation for the checkpoint assessment, including how they revised, what they revised, the time spent on revision, and what areas they could improve upon. For more strategies, see the Metacognition spread.

Intervention activity: Draw the magnetic field around a bar magnet, including the direction of the magnetic field lines.

Visualising the magnetic field

Show students the shape of the magnetic field around a bar magnet using a variety of sources. This can be using a simulation (PhET have a great interactive magnetic field to explore), a labelled diagram, and a real-life demonstration using iron filings and a bar magnet. Students should recognise the repeated shape of the magnetic field in each example.

Labelling the magnetic field

Once the shape of the magnetic field around a bar magnet is recognisable, label the direction of the field lines. Students should remember the general rule, that magnetic field lines point from 'north to south' – the lines come out of the north pole and go into the south pole. This is because field lines are drawn to represent the motion of an 'imaginary north pole', which would be repelled from other north poles and attracted to other south poles.

Ask students to assess a series of sketches of correct and incorrect magnetic fields around bar magnets. These might have, for instance, missing field lines, incorrect shapes, the field lines might not connect to the poles, or field line directions might be incorrectly labelled.

Intervention activity: Apply Fleming's left-hand rule to determine the direction of the force acting on the conductor.

Fleming's left-hand rule

Just setting up the left hand can be challenging when practising Fleming's left-hand rule. The first step is to point, with the index finger, straight ahead, then raising the thumb upwards. Model this first step for the students, asking them to follow along. It is useful to sit or stand alongside the students, so you are all facing the same direction. With the index finger pointing ahead, and the thumb directly upwards, the final step is to point the middle finger out to the right. This is tricky for many students, and be mindful of any student with a physical disability that prevents them from doing this at all. Spend time just 'setting up' their left hands in this way – repeating this to form some muscle memory.

The force on a conductor

Explain to students that Fleming's left-hand rule might seem rather strange, but it's actually a useful tool to remember the directions of three things in a motor that act at 90° to each other. This is why it's so important to have the fingers on the left hand set up in the right way.

Go through what each finger represents in turn. Use a diagram, and also the helpful pneumonic – First finger = Field, seCond finger = Current, thuMb = Motion or force.

Use plenty of very simple diagrams showing a horizontal magnetic field (i.e., towards the left and/or right) and conducting wires across these fields with current going up and/or down. Students should use the language 'into the page', or 'out of the page', to describe the direction of the force on the conductor. Plenty of practise is key.

Intervention activity: Describe how the primary current compares to the secondary current in a step-up transformer.

Visualising the output of a step-up transformer

Show students the input and output of a step-up transformer. This can be done using an oscilloscope (ideally, with the input and output trace overlaid), or by using a simulation. Lead students to observe and describe the larger amplitude (i.e., the larger peak current) of the output waveforms produced.

Explaining and practising

Recap the key ideas of the generator effect – specifically, that the number of turns (or length of conductor) in a magnetic field means that a larger p.d. (and current) can be induced. Students should link this to the output signals observed, and then practise identifying these outputs using a few examples.

P16 Space

P16.1 Formation of a star

AQA GCSE Physics specification points

4.8.1.1 Our Solar System

***Smart Curriculum* Narrative**

Earth in space

Theme/subtheme

Space/Other celestial bodies

Disciplinary knowledge

- MS 1a, 1b

Demonstration

From birth to the main sequence

	Learning outcomes		
Learning objectives	**Developing Aiming for 3**	**Secure Aiming for 5**	**Extending Aiming for 8**
Describe a variety of objects within the Solar System	Describe how the Solar System is a small part of the Milky Way galaxy	**Describe a variety of objects within the Solar System**	Analyse data about objects in the Solar System
Explain the role of gravitational forces at the start of a star's life cycle	State that fusion reactions release energy in a star	**Explain the role of gravitational forces at the start of a star's life cycle**	Explain why there has to be sufficient mass to form a star
Describe the forces that are acting when a star is in its main sequence	State that the Sun is in its main sequence and is stable	**Describe the forces that are acting when a star is in its main sequence**	Explain how a star becomes stable

Tier 2 vocabulary	Tier 3 vocabulary
distance, energy, force, fuel, helium, hydrogen, light, *mass*, pressure, radiate, star, Sun, temperature, time, wavelength	asteroid, asteroid belt, black hole, comet, fusion, gravity, main sequence, moon, nebula, nuclear, planet, protostar, pulsar, Solar System, universe

Lesson plan

Reactivate knowledge

1 Place these objects in order of increasing size: Sun, galaxy, Mars, Solar System, universe.

2 Name the eight planets in our Solar System.

3 Name three objects in our Solar System apart from the planets.

Common misconception

Students often think the Sun is burning or is a ball of fire. Respond that it does not burn but shines because of nuclear fusion, where lighter atoms are forced together to become heavier atoms, releasing enormous amounts of energy. Images of the Sun make it appear fiery, but the chemical process that creates fire (combustion) is very different to the nuclear processes happening in the Sun (fusion).

Answers: 1 Mars, Sun, Solar System, galaxy, Universe *2* Mercury, Venus, Earth, Mars, Jupiter, Saturn, Uranus, Neptune *3* Any three from: moons, Sun, satellites, asteroids, dwarf planets, comets

Trigger interest

Riding light (10 mins)

Show a video clip to visualise a trip through the Solar System. Ask students to discuss the variety of objects they observe. Alternatively, show the distances between the Sun and planets.

Challenge questions can be provided, for example: *Why do the planets orbit the Sun?* (due to gravity) *What factors affect the size of this force?* (distance and mass).

Emphasise the scale of the Solar System – light travels at 3×10^8 m/s, so some students might be able to calculate the distance between the Sun and the Earth, if they are told that it takes eight minutes to travel between them (480 seconds $\times 3 \times 10^8$ m/s = 1.44×10^{11} m).

Exposition

Solar System recap (5 mins)

Briefly recap the structure of the Solar System, outlining the definitions of planets, moons, comets and asteroids, and their basic properties. Discuss the origin of the materials that form these objects – it could be powerful to show the *James Webb Space Telescope*'s recent images showing complex asteroid belts around a star other than our own, for example, Fomalhaut.

Main activity

From birth to the main sequence (30 mins)

Ask pairs of students what our Sun is made of, collecting suggestions on mini-whiteboards. It might be useful to provide some hints for students to select, for example, hydrogen gas, helium gas, light, fire, and so on. This can open up some useful discussion around misconceptions about the Sun, and about stars in general.

Describe the formation of a protostar. Emphasise the abundance of hydrogen in the initial nebula, which acts as 'fuel' for the birth of the star. Another short video clip or animation is useful to show this process – students should be able to visualise the pull of gravity on dust and gas, which become extremely densely packed together, creating a very high pressure in the core. Discuss the increase in temperature during this first phase, which, along with the high pressure, is an ideal environment for nuclear fusion. Fusion radiates enormous amounts of energy outwards and is the reason why stars shine as part of their main sequence.

Frame the forces acting in a star as a tug-of-war between the inward pull of gravity and the outward push of radiation pressure. If either of these forces change, then the star will change size.

Finally, use the practical demonstration to discuss forces in balance, although the forces are not the same type of force.

Review and practice

Solar System objects (10 mins)

Show images of some Solar System objects and ask students to identify them, using mini-whiteboards, and describe their properties and behaviour.

Extend this activity, where appropriate, by using more obscure images, for example, images of Jupiter and Saturn in different wavelengths of light, or surface images of the Moon, Mars, or Venus.

Alternatively, students could answer 'true' or 'false' to whether a selection of images shows objects that exist in our Solar System, for example, another star, a black hole, an asteroid belt, a pulsar, and so on. This opens up opportunities to explore the diversity of objects in the universe; emphasise those that are 'local' to our home Solar System.

P16.2 The life cycle of a star

AQA GCSE Physics specification points

4.8.1.2 The life cycle of a star

Smart Curriculum Narrative

Earth in Space

Theme/subtheme

Space/Lifecycle of stars

Disciplinary knowledge

- MS 1a, 1b

Activity

The production of elements

Learning objectives	Learning outcomes		
	Developing Aiming for 3	Secure Aiming for 5	Extending Aiming for 8
Describe the life cycle of a star the size of the Sun	Identify the sequence of development for a small star such as the Sun from a diagram	**Describe the life cycle of a star the size of the Sun**	Explain the stages in the life cycle of the Sun
Describe the life cycle of a more massive star	State that changes in the fusion processes in a star result in changes in its appearance	**Describe the life cycle of a more massive star**	Explain the stages in the life cycle of a massive star
Explain how fusion leads to the formation of new elements	Describe how lighter elements are formed in stars and elements heavier than iron in a supernova	**Explain how fusion leads to the formation of new elements**	Explain why our Solar System must have formed from the remnants of a supernova

Tier 2 vocabulary	Tier 3 vocabulary
energy, gold, lead, *mass*, mercury, silver, star, surface area, temperature, *tin*, uranium, zinc	atomic number, core, electromagnetic spectrum, *element*, expansion, fusion, life cycle, main sequence, red giant, red super giant, supernova, universe

Lesson plan

Reactivate knowledge

1 What is nuclear fusion?

2 Why are main sequence stars stable?

3 What is the fuel of main sequence stars?

Common misconception

Students may think that stars live forever. Respond that stars are born, live, and then they die. This process is called the 'life cycle' of the star. What each of these phases of life and death look like depends primarily on the mass involved in the Star – for instance, the Sun will eventually become a red giant. Other stars with much more mass than the Sun will become a red supergiant and go supernova.

Answers: 1 The joining of two lighter nuclei to form a heavier nuclei *2* The outward force from nuclear fusion balances the gravitational collapse *3* Hydrogen

Trigger interest

We are all made of stardust (10 mins)

Ask students to discuss in small groups what they think is meant by the phrase *'We are all made of stardust'*. Is it a scientific statement?

Support this starter with laptops/tablets, where available, so students can research this topic and summarise what they find in two or three bullet points. Dr Maggie Aderin-Pocock, for example, could be signposted an eminent astronomer, has written informative and inspiring material on how human beings might come from the remnants of dead stars. Ask for responses, or use sticky notes to share some of the written answers with the whole group.

Exposition

The life history of a star (15 mins)

Video and/or animation is a useful tool for this part of the lesson. A common misconception may be that students think that all stars are the same. Use this lesson to explain that stars come in a large range of sizes, masses, brightnesses, and temperatures. Depending on these variables, stars will both live and die very differently.

Recap the main sequence and emphasise the expansion of the star as its core temperature rises (as more energy is released). Since the surface area becomes dramatically larger, it will cool down – ask students how this will affect the star's colour, referring back to the electromagnetic spectrum.

Describe the final stages of a small star such as the Sun, emphasising that these changes are in the very distant future. Compare these processes with the changes in a much larger star – images of a supernova outshining its galaxy are a powerful visual.

Main activity

The production of elements (20 mins)

Discuss the production of elements with reference to the Periodic Table. It might be helpful to remind students of atomic number, which increases following nuclear fusion, so that students appreciate what's happening inside a star's core right the way up to iron.

Explain that producing even heavier elements requires a net energy input, so this cannot happen in normal stars – but these heavier elements occur naturally. How can this be? Encourage students to identify that supernovae provide the extraordinary outward pressure and temperature to fuse heavy elements together. These explosions are the only place in the universe where metals such as zinc, silver, tin, gold, mercury, lead, and uranium are produced. These elements are flung into space as debris.

Refer back to the starter discussion to reflect on the presence of these metals on Earth.

For additional challenge, show students some examples of nuclear fusion equations for heavier elements.

Review and practice

Star sequence sort (10 mins)

Students use the interactive activity to match the stages of the life cycle of stars with their descriptions. They then put these stages into the correct order.

P16.3 Planets, satellites, and orbits

AQA GCSE Physics specification points

4.8.1.3 Orbital motion, natural, and artificial satellites

Smart Curriculum Narrative

Earth in space

Theme/subtheme

Space/Lifecycle of stars

Disciplinary knowledge

- MS 1a, 1b, 1c

Demonstration

Orbital forces

	Learning outcomes		
Learning objectives	**Developing Aiming for 3**	**Secure Aiming for 5**	**Extending Aiming for 8**
Compare the orbits of planets, moons, and artificial satellites	State that gravity provides the force that allows planets and satellites to maintain their orbits	**Compare the orbits of planets, moons, and artificial satellites**	Analyse data on satellite orbits and explain how the satellite could be used
(H) Explain how a satellite in a circular orbit is accelerating even when moving at a constant speed	Describe how, for an object to be moving in an orbit, there must be a gravitational force acting directed at the centre of the orbit	**(H) Explain how a satellite in a circular orbit is accelerating even when moving at a constant speed**	Explain, using Newton's laws, if the forces acting on a satellite are balanced or not
(H) Describe how if the speed of an object in orbit changes, the orbit must change	Describe that, to stay in an orbit at a particular height, the speed must stay the same	**(H) Describe how if the speed of an object in orbit changes, the orbit must change**	Explain why a geostationary satellite must be a specific distance from the centre of the Earth

Tier 2 vocabulary	Tier 3 vocabulary
air resistance, circular, force, gravity, horizontally, speed, weight	centripetal force, comet, ellipse, Halley's comet, orbit, planetary, projectile, satellite

Lesson plan

Reactivate knowledge

1 Name an object that orbits a planet.

2 Velocity is a vector quantity – what does that mean?

3 What would you need to measure to calculate the speed of an object moving in a circle?

Common misconception

Students may think planetary orbits are perfectly circular or are very elliptical. The planets in the Solar System do have elliptical orbits, which are 'flattened' circles. These are not extremely elliptical, however – most (and certainly Earth's) can be accurately modelled as circular.

Answers: 1 Moon (natural satellite) or satellite *2* Velocity has both size and direction *3* Speed = distance/time so you need to measure the time for one lap and the distance, which is circumference of the circle (calculated by $2\pi r$)

Trigger interest

Curved paths (10 mins)

Set up a thought experiment with the students, asking them to imagine carrying a projectile (e.g., a tennis ball) to the top of a hill overlooking a lake. Ask them to sketch, using mini-whiteboards, the path of the projectile if they were to throw it across the lake.

Extend this thought experiment by asking the students to discuss and suggest the forces acting on the projectile throughout its flight. You might, if needed, provide keywords (e.g., weight, gravity, air resistance). Some students can sketch the force vectors directly.

Exposition

Orbital forces (15 mins)

Refer to the sketches shown by the students, to continue the narrative about projectiles falling under gravity, drawing or showing a similar diagram that shows how, given a larger initial velocity, the object can be launched into an orbit around the Earth (e.g., Figure 2 in the Student Book).

This is an example of an artificial satellite's orbit, and students should suggest what other natural (i.e., not human-made) and artificial objects are in orbit, leading to a discussion about the Moon's orbit. Ensure the force of gravity is labelled on the diagram, and that students appreciate that an object will orbit in a circular path because this force is directed inwards (towards the centre of orbit).

Finish with a discussion about the orbits of planets in the Solar System.

Main activity

Orbital forces (20 mins)

Use the practical demonstration to show that for an object to travel in a circle (or ellipse), there must be a force acting on it, directed inwards towards the centre of orbit. This is called the centripetal force.

Following the demonstration, students can use the Bump up your Grade worksheet to develop their understanding of the centripetal force. Ensure that students understand the factors that affect the size of this force and the direction of acceleration, linking this to Newton's Second Law.

Review and practice

Halley's comet (10 mins)

Give students a brief historical introduction to the famous Halley's comet; observed periodically throughout human history, last seen in 1986 and predicted to reappear in 2061. Show students a diagram representing the elliptical orbit of the comet around the Sun, including its direction of travel and the direction of gravity, at four points on its journey.

Extend this discussion by labelling the point in the comet's orbit nearest to the Sun, and furthest away from the Sun, asking students to suggest where the comet's speed is fastest and slowest. Alternatively, students may be interested in the fact that analysis of the orbits of comets have hinted at the presence of a yet-undiscovered 10th planet – between one and ten times the size of Jupiter! This planet, if it exists, would be about five trillion kilometres out from the Sun and is (currently) invisible to telescopes.

P16.4 The expanding universe

AQA GCSE Physics specification points

4.8.2 Red-shift

Smart Curriculum Narrative

Earth in Space

Theme/subtheme

Space/Observing the universe

Disciplinary knowledge

- MS 1b, 2g

Activity

Doppler effect and red-shift

	Learning outcomes		
Learning objectives	**Developing Aiming for 3**	**Secure Aiming for 5**	**Extending Aiming for 8**
Describe how the wavelength of a wave is changed if the source of the wave is moving away from the observer	State that the wavelength of a wave is changed by the movement of the source	**Describe how the wavelength of a wave is changed if the source of the wave is moving away from the observer**	Identify red-shift or blue-shift by comparing emission spectra of objects with those of a non-moving source
Compare galaxies in terms of their red-shift and distance from us	State that a galaxy showing red-shift is moving away from us	**Compare galaxies in terms of their red-shift and distance from us**	Identify the relationship between the red-shift of a galaxy and its speed of recession from a data set or graph
Describe how red-shift provides evidence for expansion of the universe	State that all galaxies are moving away from each other and that this shows the universe is expanding	**Describe how red-shift provides evidence for expansion of the universe**	Explain how red-shift data is used to show that the universe is expanding

Tier 2 vocabulary	Tier 3 vocabulary
distance, kilometre, light, prism, spectrum, star	doppler effect, expanding, frequency, galaxy, light year, Milky Way, red-shift, *space*, spacetime, spectral line pattern, starlight, universe, wavelength

Lesson plan

Reactivate knowledge

1 What is the wavelength of a wave?

2 Which wavelength of visible light has the longest wavelength?

3 Which wavelength of visible light has the shortest wavelength?

Common misconception

Students may think the 'expanding universe' refers to galaxies moving outwards into empty, unoccupied space. Respond that galaxies *are* moving through space, but space itself is expanding too. Like an inflating balloon, the distance between two points on the surface of the balloon increases as the balloon expands. In a similar way, space (more strictly, spacetime) is constantly expanding, moving matter along with it as it does so.

Answers: 1 The distance between a point on a wave and the same point on the adjacent wave *2* Red *3* Violet

Trigger interest

What's in a galaxy? (10 mins)

Show students an image of the Milky Way galaxy. Discuss our position in the Milky Way, and the number of stars it holds (approximately 100 billion, or 100×10^9 stars).

Then, talk about the number of galaxies in the universe. These numbers are inspirational – use this opportunity to impress students with the scale of the Milky Way galaxy and the size of the universe. For example, students could guess which number is greater – the number of grains of sand on the Earth (approximately 10^{16}) or the number of stars in the universe (10^{21}).

Support this activity by using a bucket of sand to illustrate comparative powers of ten, and how each additional factor of 10 drastically increases the numbers involved. For example, in a typical bucketful, there are about 300 million grains of sand. This is 3×10^8 grains. 3×10^9 represents another 10 buckets, and 3×10^{10} represents another 100 buckets!

Exposition

Starlight (15 mins)

Explain that studying the light emitted from stars in galaxies tells scientists lots of information. For instance, passing the light received on Earth through a prism reveals the spectrum of colours (wavelengths) produced by the stars in the galaxies. Show a spectral line pattern from the Sun, identifying the positions of key dark lines in the spectrum, and then compare this to a spectrum of an object moving away. Students should identify that these lines 'shift' towards the red end of the spectrum.

Main activity

Doppler effect and red-shift (20 mins)

Show students a video clip of a passing police car or ambulance. Students should identify the changes in pitch (frequency) – this is a good everyday example of the doppler effect.

Explain that wavelengths are increased, or stretched out, if the object moves away from the observer. These shifts become larger when the objects are moving away more quickly – link this back to the idea about spectral lines, and the fact that spectral lines from distant galaxies are very red shifted, to conclude these galaxies are moving away from the Earth.

Show students an image of the universe from the *James Webb Space Telescope*, pointing out a variety of galaxies and discussing the relative distances – some are much closer than others. Discuss the surprising fact that the further galaxies are away from the Earth, the more their light is red-shifted – and hence, the faster they are moving away from the Earth. Show this using a recession graph for galaxies, clearly referencing the linear proportionality between distance and recession speed.

Extend this part of the discussion by asking the students what this graph implies about how the universe evolved, that is, how galaxies were distributed in the distant past. It helps to ask students to imagine 'rewinding' a video recording of all galaxies in the universe, given that right now, they seem to be moving away from each other. Lead them toward the idea that, long ago, they were much closer together, and the universe was much smaller (and has been expanding ever since).

Review and practice

Space is big (10 mins)

Students can use the interactive to put a list of distances in order (e.g., the school to Paris, the school to New York, the Earth to the Moon, the Earth to the Sun, etc.). They can then match these to actual distances in kilometres.

Extend students by asking them to convert the numbers into standard form, and discuss why this is generally used to represent large numbers.

P16.5 The beginning and future of the universe

AQA GCSE Physics specification points
4.8.2 Red-shift

Smart Curriculum Narrative
Earth in Space

Theme/subtheme
Space/Theories of the universe

Disciplinary knowledge
- WS 1.1, 1.2, 1.3
- MS 1b

Demonstration
The Big Bang model

Learning objectives	Learning outcomes: Developing Aiming for 3	Learning outcomes: Secure Aiming for 5	Learning outcomes: Extending Aiming for 8
Describe the Big Bang theory	State that the currently accepted model for the early universe is the Big Bang model	**Describe the Big Bang theory**	Explain why measurements from Earth showing the universe is expanding do not mean the Earth is the centre of the universe
Describe the evidence for the Big Bang theory	Describe how scientists look for new evidence all the time	**Describe the evidence for the Big Bang theory**	Outline recent discoveries that have led to changes in the theories of how the universe will develop
Explain that there is much still to understand about the universe	State that scientists are looking into new ideas to understand the universe such as dark matter and dark energy	**Explain that there is much still to understand about the universe**	Discuss how scientists use new evidence and change their theories about how the universe has evolved over time and how it will change in the future

Tier 2 vocabulary	Tier 3 vocabulary
energy, *mass*	Big Bang, cosmology, dark matter, dense, galaxy, microwave, radiation, red-shift, spacetime, universe, wavelength

Lesson plan

Reactivate knowledge

1. What is red-shift?
2. What does it mean if light is more red-shifted?
3. Describe the relationship between the speed of a galaxy and its distance from the Earth.

Common misconception

Students often think the Big Bang is the explosion that began our universe. Clarify there was no explosion – rather, the term 'Big Bang' refers to the extremely rapid expansion of spacetime at the very early stages of the universe.

Answers: 1 The increase in wavelength from light from distant galaxies *2* The galaxy is moving faster *3* The greater the distance the greater the speed

Trigger interest

Ultimate timeline (10 mins)

Show students an image of the evolution timeline for the universe, which physicists have developed according to observations of how the universe looks now, including observations of the 'leftover' microwave radiation from an event called the Big Bang. Touch on a few key events, for example, the start of the universe (13.8 billion years ago), the formation of the Solar System (4.6 billion years ago), and first life on Earth (3.6 billion years ago). It is important that the Big Bang theory is framed as a model for how the universe has evolved, and like all scientific models, should continue to be tested and refined according to new scientific evidence.

Exposition

The Big Bang model (25 mins)

Students will need to be able to describe complex processes, so providing them with gap-fill style exercises or summaries, to highlight or underline key words or phrases, is useful. Comprehension activities work well, especially if paired with appropriate video clips or animations of Big Bang simulations.

Recap the idea that all the observable universe was once in the same place many billions of years ago, based on observations of red-shifted spectral lines. Describe the extremely hot, dense conditions that would have existed at that time, because all the mass and energy contained in the universe would have been concentrated in an unimaginably small region.

Demonstrate expansion using the practical activity and outline the changes that occur during expansion, that is, galaxies moving apart. Hence the universe cooling and becoming less dense.

After demonstrating expansion, ask students to suggest the things that would be affected by this expanding, or stretching process. Lead them to the idea that even electromagnetic radiation, which is emitted by all hot objects, would have been stretched (or increasing in wavelength). The remnant of this process is the cosmic background radiation, observable today as a uniform spread of faint radiation in the universe. Discuss how the Big Bang theory is also supported by the observation that such red-shifted radiation is observed from most galaxies, suggesting expansion in all directions.

Main activity

The Big Bang model (10 mins)

In this demonstration, a balloon can be inflated to show that the galaxies drawn on its surface move away from each other when space (the rubber) stretches. This is similar to the expansion of the universe, but the balloon is expanding in two (curved) dimensions whereas the universe expands in four dimensions – three of space and one of time.

Review and practice

Sceptic (10 mins)

Pose students the following questions:

- Are there competing scientific observations that challenge the Big Bang model?
- Has the *James Webb Space Telescope*, or other scientific instrument or theory, predicted or discovered unexplained or contradictory evidence that modern cosmology cannot answer?

Students can use the interactive to summarise the current evidence for the expanding universe and discuss the unresolved questions that remain – i.e., the presence of dark matter, and the ultimate fate of the universe.

Introduction to checkpoint intervention

The checkpoint intervention provides suggestions for a lesson to follow up the *P16 Space* checkpoint assessment. Depending on the outcomes of the assessment, these suggestions can be used to move a class towards secure, to target specific learning outcomes, or to consolidate knowledge for students achieving secure outcomes. Use students' outcomes from the checkpoint assessment to address general misconceptions from the content covered.
Be prepared to re-cover content as required; three key concepts from the chapter are covered in more detail below.

Checkpoint secure learning outcomes

- Explain the role of gravitational forces at the start of a star's life cycle.
- Explain how fusion leads to the formation of new elements.
- (H) Describe how, if the speed of an object in orbit changes, the orbit must change.
- (H) Explain how a satellite in a circular orbit is accelerating even when moving at a constant speed.
- Compare galaxies in terms of their red-shift and distance from us.
- Describe the Big Bang model.

Student reflection

Encourage students to reflect on whether there were any checkpoint questions they found difficult or straightforward as well as their preparation for the checkpoint assessment, including how they revised, what they revised, the time spent on revision, and what areas they could improve upon. For more strategies, see the Metacognition spread.

Intervention activity: Explain the role of gravitational forces at the start of a star's life cycle.

Create a set of true or false statements about the properties of gravity. Students can answer these on mini-whiteboards to gain confidence in recognising key facts about how gravity behaves.

For instance: it is an attractive force, it exists between all objects in the universe with mass (including 'light' objects like rocks and grains of dust), and the size of the gravitational force increases with the amount of mass involved.

Since gravity is an attractive force, the particles that make up rocks, gas, and dust in space are able to pull on each other, speeding up as they get closer together. In this way, clumps of matter gradually get larger until a protostar is formed, which is a 'star-to-be'. It is effective to describe this process verbally, alongside a visual (e.g., an animation or video clip), after which students can summarise the key ideas in short bullet points.

Extend the discussion to the transfer of energy between the protostar's gravitational potential and thermal energy stores, as the particles get closer together, which leads to high enough temperatures to ignite fusion.

Ⓗ Intervention activity: Explain how a satellite in a circular orbit is accelerating even when moving at a constant speed.

Remind students that vector quantities have magnitude and direction. Provide a card-sorting activity for them to identify and sort vectors from scalars, including speed, velocity, and acceleration. Extend this by including some cards with values and units, e.g., +5.0 N, –2.1 N, +4.0 m/s, –9.5 m/s, and –10.4 m/s^2. Discuss the importance of the +/– signs, which indicate that the quantity is directional (and is a vector).

Show students the Ladybug PhET simulation, or similar, of an object in circular motion, with the changing velocity vector displayed. Alternatively, use an image of a satellite in orbit around the Earth, asking students to draw the direction of motion (the velocity vector), at four points along the satellite's orbit. Does the direction of motion change? Convince students using this visual that since the direction is changing, then the velocity must be changing, because velocity (being a vector) is made up of both magnitude and direction. The students can then attempt to justify this further, given the speed of the orbiting satellite is constant, for example, annotate a constant speed at each of the four points along its orbit, or ensure the animation used makes this clear.

Intervention activity: Describe the Big Bang model.

There are a variety of ways for students to interact with the story of the Big Bang. For example, they could be provided with a short bullet-point summary of the Big Bang model, and a set of images. Using these, they could identify and match the corresponding image for each point. Alternatively, these bullet points could be shuffled so the final stage would require them to arrange sentences and images chronologically. If needed, give timeline hints alongside each image and/or sentence (e.g., 13.8 billion years ago, 4.5 billion years ago, etc).

Since this intervention requires a fair amount of cognitive recall, providing students with frameworks wherever possible will be useful. Gap-fill exercises to summarise their learning, using keywords, is a short and effective way to assess their learning here.

Answers

P1.1

1a Energy transfer from gravitational to kinetic store [1] some energy transferred to thermal store of surroundings [1]

b Electric current transfers energy to thermal store of heating element [1] energy transferred by heating to thermal store of surroundings [1]

2a Any two from: e.g., electric torch: chemical energy stored in battery → electric current transfers energy to lamp → energy transfer to thermal store of surroundings by light and by heating [2] e.g., candle: energy from chemical reactions when candle burns → energy transferred to thermal store of surroundings by light and by heating [2]

b i Candle [1]

ii Electric torch [1]

3a Energy transferred by electric current to train motor [1] → energy transferred to gravitational store of train [1]

b Friction on train wheels due to brakes reduces energy in kinetic store of train to zero [1] → energy transferred to thermal store of brakes and to surroundings by heating and sound waves [1]

4 Electrical energy supplied to oven used to generate microwaves, [1] microwaves transfer energy to thermal store of the food → [1]

P1.2

1a Friction transferred energy from kinetic store [1] energy transferred by heating to thermal store of brakes and surroundings [1]

b Up: energy transferred from kinetic to gravitational store [1] Down: Energy transferred from gravitational to kinetic store [1] throughout energy transferred to thermal store of surrounding by friction/air resistance [1]

2a Descent: Energy transferred from gravitational store to kinetic and thermal store of the surroundings due to air resistance, [1] impact: Energy from kinetic to elastic and thermal store of trampoline, [1] ascent: energy from elastic store to kinetic to gravitational store and thermal store of surroundings due to air resistance [1]

b Less energy in gravitational store at top of bounce than at point of release [1]

c Clamp metre ruler vertically over middle of trampoline, hold ball next to ruler with lowest point level with top of ruler, [1] release ball and observe highest level of bottom of ball after rebound, [1] repeat several times and calculate average rebound position, [1] repeat with same ball for other two trampolines, [1] highest rebound position is the bounciest [1]

3 Energy in elastic store of rubber straps → KE of capsule, [1] KE → GPE of capsule [1]

P1.3

1a i Energy transferred from chemical store of muscles to kinetic store of boat and thermal store of surroundings [1]

ii Energy transferred from kinetic store of motor to gravitational store of barrier to thermal store of surroundings by friction and sound waves [1]

b 3000 N × 50 m = 150 000 J [1]

2a Friction between brake pads and wheel discs do work [1] energy is transferred from kinetic store of car to thermal store of brakes [1]

b 7000 N × 20 m = 140 000 J [1]

3a i 20 N × 4.8 m = 96 J [1]

ii 80 N × 1.2 m = 96 J [1]

b $\frac{1400\text{ J}}{7.0\text{ m}}$ [1] = 200 N [1]

P1.4

1a Descent: energy transferred from gravitational store by gravity to kinetic store [1] impact: energy transferred from kinetic to elastic store to kinetic store [1] after impact energy transferred from kinetic to gravitational store [1] energy transferred to thermal store of surroundings due to work done by air resistance throughout [1]

b i mg = 1.4 N = 1.4 N × (2.5 m − 1.7 m) [1] = 1.1 J [1]

ii Any two from: energy → surroundings due to air resistance as ball moves through air, [1] energy transfer due to heating of ball when it is deformed, [1] energy → surroundings by sound waves when ball hits floor [1]

2a 550 N × 0.20 m = 110 J [1]

b 50 × 110 J = 5500 J [1]

3a 25 kg × 9.8 N/kg × 1.8 m [1] = 441 J [1]

b Ball falls 1.8 m − 0.3 m = 1.5 m [1]

$E_p = mgh$ = 25 kg × 9.8 N/kg × 1.5 m = 367.5 J [1]

4 Energy transferred from chemical store of muscles to keep them contracted [1] no work done on object as it doesn't move, [1] energy transferred to thermal store of muscles and by heating to thermal store of surroundings [1]

P1.5

1a 0.5 × 500 kg × (12 m/s)^2 [1] = 36 000 J [1]

b 0.5 × 0.44 kg × (20 m/s)^2 [1] = 88 J [1]

2 Elastic energy $E_e = \frac{1}{2}ke^2$ = 0.5 × 250 N/m × (0.21 m)^2 [1] = 5.5 J [1]

3a i Work done by muscles transfer energy from chemical store of muscles [1] → elastic PE of catapult [1]

ii Elastic PE of catapult [1] → KE of object [1]

b i $E_e = \frac{1}{2} \times 80 \times 0.5^2 = 10\text{ J}$

ii Mass = weight/g = 2.0 N ÷ 10 N/kg = 0.20 kg [1] assume all elastic energy → KE in object, $E_K = \frac{1}{2}mv^2 = 10\text{ J}$

so $\frac{1}{2} \times 0.20\text{ kg} \times v^2 = 10\text{ J}$ [1] $\therefore v^2 = \frac{10\text{ J}}{0.5 \times 0.20\text{ kg}}$ [1]

$= 100\text{ m}^2/\text{s}^2 \therefore v = 10\text{ m/s}$ [1]

4a Work done by brakes = energy transferred from kinetic store = 360 000 J [1] using $W = Fs$ → 360 000 J = F × 100 m [1]

$\therefore F = \frac{360\,000\text{ J}}{100\text{ m}}$ = 3600 N [1]

b $\frac{1}{2} \times m \times (30\text{ m/s})^2 = 360\,000\text{ J}$ [1] $\therefore 0.5 \times m \times 900\text{ (m/s)}^2 =$ 360 000 J, rearranging → $m = \frac{360\,000\text{ J}}{0.5 \times 900\text{ m}^2/\text{s}^2}$ [1] = 800 kg [1]

P1.6

1a Wasted: sound [1]

b Useful: light and sound, wasted: heating thermal store of TV surroundings [1]

c Useful: energy transferred to thermal store of water wasted: energy transfer to thermal store of surroundings by heating and sound [1]

d Useful: sound, wasted: energy transfer to thermal store of surroundings [1]

2a Gear box would heat up due to energy transfer through friction between the gears, [1] the hotter the gear box, the less efficient the gears, [1] if gear box becomes very hot, stops working as oil in it burns up and gear wheels wear away [1]

b Inside of shoes heat up due to energy transfer (by conduction and infrared radiation) from feet which rub, [1] feet transfer less energy as shoes warm up so feet and athlete become hotter [1]

c Drill heats up due to friction between rotating drill and wood, [1] if drill becomes very hot, it burns wood creating smoke [1]

d Discs heat up due to energy transfer by friction between discs and brake pads, [1] KE of car decreases [1]

3a As pendulum swings towards middle energy transferred from gravitational to kinetic [1] as it moves from middle to highest position energy transferred from kinetic to gravitational store [1] throughout energy transferred by work done by air resistance [1] to thermal store of surroundings [1]

b As air resistance opposes motion work done transfers energy from kinetic to thermal store of surroundings, [1] until pendulum stops moving and has no KE, [1] energy transferred to air dissipated to surroundings [1]

4 Friction at wheel axles [1] and air resistance, [1] work done transfers energy from kinetic store to thermal store of surroundings [1] sound may also transfer energy to surroundings [1]

P1.7

1a Useful energy < energy supplied [1]

b i Efficiency = useful energy delivered/input energy supplied, [1] a machine never > 100% efficient because useful energy delivered never > total energy supplied/because of conservation of energy [1]

ii Useful energy always < total energy supplied, [1] due to electric currents in circuit wires/components [1] and friction between moving parts, [1] energy dissipated by transferring heat to surroundings [1]

2a 60 J − 24 J = 36 J [1]

b $\frac{24\,\text{J}}{60\,\text{J}}$ (× 100%) [1] = 0.40 (or 40%) [1]

3 0.25 × total energy supplied [1] = 0.25 × 3200 J = 800 J [1]

4 Electric current transferred energy to fan heater → energy in thermal store of heater and kinetic store of fan increases [1] energy in thermal store of air increases by heating and by work done by fan [1] energy wasted because sound waves transfer energy to surroundings [1] and work done by friction between moving parts transfers energy to thermal store of moving parts [1]

P1.8

1a **Y** [1]

b **Y** [1]

c **X** [1]

2a Work done winding clockwork spring up [1] to increase energy in elastic potential store of spring [1]

b Work done by spring transfers energy from elastic store [1] to the generator, electric current transfers energy to the radio circuits [1] energy transferred to the surroundings by sound [1] and by heating to the thermal store of surroundings [1]

c Advantage: no replacement batteries needed [1] disadvantage: spring needs to be rewound after unwinding [1]

3a Heating water, [1] energy transferred to kinetic store of motor to pump and spray [1]

b Hot water transfers energy by heating to surroundings, [1] machine vibrations create sound waves → energy to thermal store of surroundings [1]

4a 80% × 60% [1] = 48% [1]

b (100% − 48%) = 52% [1]

P1.9

1a i Mains filament bulb [1]

ii 10 000 W electric cooker [1]

b 2 million × 3 kW [1] = 6 million kW [1]

2a 5000 W × 20 s [1] = 100 000 J [1]

b 12 000 J (× 100%)/100 000 J [1] = 0.12 (or 12%) [1]

3a $\frac{1500\,\text{kJ}}{50\,\text{s}}$ [1] = 30 kW [1]

b $\frac{30\,\text{kW}}{100\,\text{kW}}$ [1] = 0.30 (or 30%) [1]

4a Energy transfer by current to heater increasing energy in thermal store of heater [1] energy transfer by heating to the water increasing energy in thermal store of water [1] energy transfer by current to kinetic store of pump [1] work done by pump increasing kinetic store of water [1] energy transfer to thermal store of surroundings by heating/sound [1]

b 12 000 W × 4 × 20 × 60 s [1] = 960 000 J [1]

P2.1

1a Any two from: do not conduct by heating, [1] so handle not hot when pan hot, [1] steel handle becomes as hot as pan as steel is good conductor [1]

b Felt: contains fibres that trap layers of air, [1] dry air good insulator [1]

2a **Z** is the best conductor [1] because the butter melted the fastest [1]

b **X Y Z** [1]

c Any two from: to check if the experiment is repeatable/to identify anomalies/to reduce the effect of random errors by calculating a mean [2]

3 Any five from: fill beaker with measured volume hot water, measure initial temperature of water, place lid on beaker and measure temperature again after 300 s, [1] stir water before measuring temperature, [1] repeat test using two identical beakers, one inside the other, using same volume hot water at same initial temperature, [1] repeat test using three or more beakers, [1] should find temperature falls less the more beakers you use, [1] plot graph of temperature decrease *vs* number of beakers used [1]

P2.2

1 Small bucket [1] because mass of water much less than in large bucket [1]

2a Lead has lower specific heat capacity, [1] for equal masses, less energy needed for a given temperature rise [1]

b i 0.20 kg × 900 J/kg °C × (40 – 15) °C [1] = 4500 J [1]

ii 0.40 kg × 4200 J/kg °C × (40 – 15) °C [1] = 42 000 J [1]

3a 30 000 = 0.5 × c × (33 – 20) [1]

$c = \dfrac{30\,000}{0.5 \times 13}$ = 4620 [1] J/kg °C [1]

b This is higher than the true value. [1] This is because energy transferred by the heater also warmed the container and the surroundings/not all the energy was transferred to the water. [1]

4 Measure mass *m* of empty beaker, fill two-thirds full with oil and measure mass of beaker and oil, calculate mass of oil, in beaker, [1] insulate beaker and use thermometer to measure initial temperature of oil, record initial joulemeter reading, [1] use heater and joulemeter to heat oil for 100 seconds then switch off power supply, [1] record final joulemeter reading, stir oil and measure its highest temperature, [1] calculate ΔE from the two joulemeter readings and temperature increase $\Delta\theta$ from the two temperature readings, use equation $\Delta E = mc\Delta\theta$ to calculate specific heat capacity of oil, [1] assuming negligible energy to heat polythene beaker [1]

P2.3

1a Conducts much less energy than air, especially if air damp, [1] prevents energy transfer by radiation (and convection) across cavity [1]

b Foil reflects infrared radiation from radiator [1] preventing absorption of radiation by surface of wall behind foil [1]

2 Energy transferred depends on the mass and specific heat capacity, [1] more energy is stored for the same temperature rise [1]

3 Wider gap transfers less energy, [1] air between panes insulates and the wider the air gap, the more effective the window is at reducing energy transfer through it [1]

4a 6 × £15 for the rolls + £90 to fit the insulation [1] = £180 [1]

b 6 × £10 = £60 [1]

P3.1

1a Coal, oil, and gas-fired power stations [1]

b Nuclear power station [1]

c Nuclear fuel produces radioactive waste, [1] which must be stored for many years until it becomes non-radioactive. [1]

2a Any three from: Coal, oil and gas increased and still largest share/oil increased the most/nuclear began after 1950s and increased then remained steady/hydro power largest share of renewables/from 1990s other renewables introduced and increasing [3]

b In remote areas traditional fuels have always been used and continue to be used. [1]

3a Biofuel is any fuel obtained from living or recent organisms, [1] ethanol is a biofuel because obtained from fermented sugar cane [1]

b CO_2 released when burnt is equal to the C taken in as CO_2 from atmosphere [1] when it grows [1]

4 Energy per person per year = 500 million million million Joules per 800 million [1] = 62 500 MJ [1]

energy per person per second = $\dfrac{62\,500\text{ MJ}}{365 \times 24 \times 60 \times 60}$ [1]

= 2000 J/s [1]

P3.2

1a Source of energy that can be replaced

b i Tidal power [1]

ii Wind power [1]

2a i 1000 [1]

ii 25 km [1]

b From top: hilly or coastal areas, estuaries, coastline, mountain areas (4 = [2] 3 = [1])

3a Tidal: sea water flows through turbines in barriers built across estuaries, [1] hydroelectric: involve less construction because uses rainwater trapped in upland reservoirs [1] **or** tidal: sea water trapped by tidal flow in estuary by long barrier, periodic not constant, hydroelectric use water flowing continuously from upland reservoirs, [1] hydroelectric is continuous, tidal power is produced for only part of each tidal cycle [1]

b i Hydroelectricity [1]

ii Only possible where hilly, not flat [1] with significant rainfall, not dry [1]

4a HEP station that uses electricity from other power stations at off-peak time to pump water to upland reservoir, [1] when demand high, flow reversed and water in reservoir used to generate electricity [1]

b Coal, oil, and nuclear power stations run continuously as cannot be restarted quickly if demand rises suddenly, [1] water in reservoirs generates electricity when demand high [1]

P3.3

1a Energy released by radioactive substances deep underground and energy from formation of the earth [1]

b Solar energy not available at night whereas geothermal energy released all the time **or** output of solar panel reduced by cloud cover whereas geothermal energy unaffected [1]

2a 300 W/0.2 W per cell [1] = 1500 cells [1]

b To supply electricity when dark [1]

3a 200 000/160 = 1250 W [1]

b Advantage: geothermal energy does not vary whereas wind energy depends on weather conditions, [1] disadvantage: geothermal power stations only operate where flow of geothermal energy from within the Earth is significant [1]

4a 0.010 kg/s × 4200 J/kg °C × (35 – 14) °C [1] = 880 J/s [1]

b 880 J/s = 0.017 kg/s × 4200 J/kg °C × $\Delta\theta$ [1]

$\therefore \Delta\theta = \dfrac{880\text{ J/s}}{0.017\text{ kg/s} \times 4200\text{ J/kg °C}}$ [1] = 30 °C [1]

P3.4

1 Coal: **B** [1]

Hydroelectricity: **D** [1]

Uranium: **C** [1]

Wind power: **A** [1]

2a Gas [1]

b Increase of CO_2 in atmosphere, [1] acid rain [1]

c Advantages: any two from: never run out, [1] do not release greenhouse gases/CO_2 into atmosphere, [1] do not produce radioactive waste, [1] disadvantages: take up large areas, [1] affect habitats of plants and animals [1]

3a Solar, [1] wave energy, [1] wind [1]

b Any three from: nuclear, geothermal, tidal, solar furnace, hydroelectric, wind farm [3]

4 Each power station requires an advantage and a disadvantage in terms of reliability, and environmental effects e.g., the use of land, effect on natural habitats, pollution, waste [5]

P3.5

1a Gas-fired [1]

b Geothermal/hydroelectric/tidal [1]

c Wind, solar, wave [1]

d Hydroelectric [1]

2 Not enough electricity at night if no wind or waves [1] more pumped storage schemes or batteries needed to store surplus electricity [1]

3 Output power not increased quickly enough to meet sudden variations in demand [1]

4 Any two from: electricity is used to pump the water to the top reservoir increasing gravitational energy [1] The water can be released at times of high demand [1] this is a way of storing excess energy produced by power stations [1]

5a Solar [1]

b Much harder to build and maintain wind turbines in the sea [1]

c Any two from: efficiency improving [1] technology/design improving [1] cheaper to construct [1] fuel prices may vary e.g. as gas reserves decrease [1] may operate for longer/less lifetime than expected [1]

P4.1

1a i Electrons transfer from cloth to polythene rod [1] electrons have a negative charge [1] so rod now has more negative charges than positive [1]

ii Electrons transfer from perspex rod to cloth [1] electrons are negatively charged [1] so there rod now has more positive charges than negative ones [1]

b Electrons negatively charged, [1] glass loses electrons so gains positive charge [1]

2 At least four arrows evenly spaced perpendicular to the sphere [1] arrows pointing to the centre of the sphere [1]

3a Attract [1]

b Attract [1]

c Repel [1]

4a **X** and **Y** have same type of charge [1]

b Suspend **R** horizontally, rub with dry cloth to charge, charge **X** and hold near **R**, [1] if **X** repels **R**, **X** also negative, [1] if **X** attracts **R**, **X** positive, [1] **Y** repels **X** so same charge as **X** [1]

5a Friction between soles of shoes and carpet causes them and their shoes to become charged, [1] touching metal radiator gives shock because radiator earthed and they are charged at very high voltage, [1] electrons transfer between them and radiator as a spark [1]

b As you move in seat clothing rubs against car seat fabric so you become charged, [1] if metal frame of car earthed, [1] when you get out of seat you get electric shock when touching metal frame of car because charge on you creates a spark between you and car frame [1]

P4.2

1 1 = cell, 2 = switch, 3 = indicator, 4 = fuse (all correct = [2], 3 correct = [1])

2a Circuit correct: diode at 2 with arrow pointing to 3 [1]

b Variable resistor [1]

c $0.25\,\text{A} \times 60\,\text{s}$ [1] $= 15\,\text{C}$ [1]

3a $3.5 \times 2 \times 60$ [1] $= 420\,\text{C}$ [1]

b $150 = I \times 0.005\,\text{s}$ [1]

$I = 150/0.005$ [1] $= 30\,000\,\text{A}$ [1]

4a Circuit with bulb, wires, and cell (all correct = [2] one component incorrect = [1])

b Electron passing through battery gains energy from chemical reactions in battery, [1] electron transfers energy to filament bulb by colliding with atoms in filament as it passes through, [1] transfers some energy to atoms in wire in same way [1]

P4.3

1 **W** $6.0\,\Omega$ [1] **X** $80\,\text{V}$ [1] **Y** $2.0\,\text{A}$ [1]

2a $\frac{9.0}{0.6} = 15\,\Omega$ [1]

b Ammeter reading increases, [1] because resistance of thermistor decreases, [1] so total resistance decreases [1]

3 If LDR covered current decreases [1] as LDR resistance increases [1] and p.d. still 9.0 V [1]

4a $\frac{12.0\,\text{V}}{0.015\,\text{A}} = 800\,\Omega$ [1]

b i $0.015\,\text{A} \times 1200\,\text{s} = 18\,\text{C}$ [1]

ii $12 = \frac{E}{18}$ [1]

$E = 12 \times 18 = 216\,\text{J}$ [1]

P4.4

1a Resistor [1]

b Diode [1]

c Filament bulb [1]

2 Choose any pair of points from the graph, e.g., $V = 2\,\text{V}$, $I = 0.2\,\text{A}$ [1]

$R = \frac{V}{I} = \frac{2}{0.2} = 10\,\Omega$ [1]

3a $\frac{0.5\,\text{V}}{0.1\,\text{A}} = 5\,\Omega$ [1]

b $\frac{2.0\,\text{V}}{0.2\,\text{A}} = 10\,\Omega$ [1]

4a Current = 0 until p.d. ≈ 0.7 V [1] then increases rapidly [1]

b Resistance very large until p.d. ≈ 0.7 V [1] then decreases rapidly [1]

P4.5

1a $1.2\,\text{V} - 0.8\,\text{V}$ [1] $= 0.4\,\text{V}$ [1]

b $5.0 = 1.0/I$ [1]

$I = \frac{1.0\,\text{V}}{5.0\,\Omega} = 0.20\,\text{A}$ [1], p.d. $= 1.5\,\text{V} - 1.0\,\text{V} = 0.5\,\text{V}$ [1]

2a Circuit with cell and 2 resistors in series [1]

b i $3.0\,\Omega + 2.0\,\Omega = 5.0\,\Omega$ [1]

ii $5.0 = 1.5.5/I$ [1] $\frac{1.5\,\text{V}}{5.0\,\Omega}$ [1] $= 0.3\,\text{A}$ [1]

c Total $R = \frac{1.5\,\text{V}}{0.25\,\text{A}} = 6.0\,\Omega$ [1] $R_x = 6.0\,\Omega - 2.0\,\Omega = 4.0\,\Omega$ [1]

3a i $2\,\Omega + 10\,\Omega = 12\,\Omega$ [1]

ii $2 \times 1.5\,\text{V} = 3.0\,\text{V}$ [1]

b $\frac{3.0\,\text{V}}{12\,\Omega} = 0.25\,\text{A}$ [1]

c $V_P = 0.25\,\text{A} \times 2\,\Omega = 0.5\,\text{V}$ [1] $V_Q = 0.25\,\text{A} \times 10\,\Omega = 2.5\,\text{V}$ [1]

4 Any four from: same current through each resistor, [1] with additional resistor in series more resistors share total p.d., [1] so p.d. across each resistor less, [1] current through resistors less, [1] total p.d. unchanged so total resistance (= total p.d. ÷ current) > before [1]

P4.6

1a 0.40 A – 0.10 A = 0.30 A [1]

b 3 Ω resistor [1]

c Battery current = 10 A ∴ current same if R of single resistor = $\frac{6.0\,\text{V}}{10\,\text{A}}$ [1] = 0.60 Ω [1]

2a Circuit diagram: 6.0 V battery across 12 Ω and 24 Ω resistors in parallel [1]

b i Current = $\frac{6.0\,\text{V}}{12\,\Omega}$ = 0.50 A [1]

ii Current = $\frac{6.0\,\text{V}}{24\,\Omega}$ = 0.25 A [1]

c Cell current = 0.5 A + 0.25 A [1] = 0.75 A [1]

3a i $I_1 = \frac{6.0\,\text{V}}{2\,\Omega} = 3.0\,\text{A}$ [1] $I_2 = \frac{6.0\,\text{V}}{3\,\Omega} = 2.0\,\text{A}$ [1] $I_3 = \frac{6.0\,\text{V}}{6\,\Omega} = 1.0\,\text{A}$ [1]

ii Total current = 3.0 + 2.0 + 1.0 = 6.0 A [1]

4 I through 2 Ω resistor = 3.0 A [1] total I = sum of currents in individual resistors so > 3.0 A [1] equivalent R = total p.d. across resistors (i.e. 6V) ÷ total current [1] as total current > 3.0 A, equivalent R < 2 Ω [1] (**or** current through 2 Ω resistor < total current [1] as all resistors contribute to total current, [1] p.d. across 2 Ω resistor = p.d. across combination, [1] so equivalent R < 2 Ω as total current > current through 2 Ω resistor and p.d. is the same [1])

P5.1

1 In this order – lose 1 mark for each incorrect: alternating, 50, live, neutral, 230, 0 [max 3]

2a longer [1] and thinner [1]

b shorter [1] and wider [1]

3 Step up transformers increase the p.d.[1] This decreases the current [1] this reduces energy loss due to heating in the wires [1]

4a D.c. in one direction only, a.c. repeatedly reverses direction [1]

b Diode only allows current in one direction, (so it rectifies a.c. to d.c.) [1]

c i Similar shape to P5.1, Figure 1 but when negative, current = 0 [1]

ii Peaks not as high [1] horizontal spacing unchanged [1]

P5.2

1a Live = brown, neutral = blue, earth = yellow and green [1]

b i So each appliance can be switched on and off [1] without affecting others [1]

ii Brass doesn't oxidise but copper does, [1] brass harder than copper and doesn't deform as easily [1]

iii Live wire could be exposed where cable is worn or damaged [1]

2 The live wire is at 230 V [1] if it is touched there will be 230 V between the person and earth [1] they will be electrocuted [1]

3a Each must be insulated to avoid dangerously large current in cable [1] due to very low resistance between live and other wires where they touch [1]

b Earth is connected to terminal fixed to metal case, [1] other end of earth wire connected to earth pin, [1] so when plug connected to wall socket, metal case is connected via earth wire to the ground [1]

4a Wall sockets: cables thicker so resistance lower and more current passes through them than through lighting cables [1] otherwise heating effect of current greater and cables would overheat [1]

b Two-core cable has only live and neutral, three-core also has earth wire [1]

c Any appliance with double-insulated plastic case can have two-core cable, [1] appliance with metal case must have three-core cable so metal case is earthed [1]

P5.3

1 5 A × 230 V [1] = 1150 W [1]

2a i 5 A × 12 V [1] = 60 W [1]

ii 12 A × 230 V [1] = 2760 W [1]

b i Current is (50 W/12 V) = 4.2 A [1] so 5 A fuse should be used [1]

ii Current is (800 W/230 V) = 3.5 A [1] so 5A fuse should be used [1]

3a In normal operation, current in oven = 800/230 = 3.5 A [1] so 3 A fuse would melt [1]

b i $\frac{12\,\text{V}}{4.0\,\Omega}$ [1] = 3.0 A [1]

ii 3.0 A × 12 V [1] = 36 W [1]

iii 36 W × 1200 s [1] = 43 200 J [1]

4a $26^2 \times 0.25\,\Omega = 169\,\text{W}$

b $\frac{169\,\text{W}}{6000\,\text{W}} \times 100\%$ [1] = 2.8% [1]

P5.4

1a 30 C × 4 V [1] = 120 J [1]

b 60 C × 12 V [1] = 720 J [1]

2a 20 C × 6 V [1] = 120 J [1]

b 3 A × 20 s × 5 V [1] = 300 J [1]

3a $\text{I} = \text{PV} = \frac{20}{12}$ [1] = 1.67A [1]

b $\frac{(1.67 \times 12)}{230}$ [1] = 0.087A [1]

4a 4.0 Ω + 8.0 Ω [1] = 12.0 Ω [1]

b $\frac{6.0\,\text{V}}{12.0\,\Omega}$ [1] = 0.50 A [1]

c 4 Ω: 2.0 V (= 0.50 A × 4.0 Ω) [1] 8.0 Ω: 4.0 V (= 0.50 A × 8.0 Ω) [1]

d 4 Ω: 60 J (= 30 C × 2.0 V) [1] 8.0 Ω: 120 J (= 30 × 4.0 V) [1]

e 60 J + 120 J [1] = 180 J [1]

P5.5

1a 5 W × 3000 s [1] = 15 kJ [1]

b 100 W × (24 × 60 × 60) s [1] = 8.64 MJ [1]

2a i 80 J (= 20% of 100 J usefully transferred = 100 – 20) [1]

ii 5 J (80% of 25 J usefully transferred = 20 J) [1]

b Efficiencies very different, efficiency of each LED much > efficiency of halogen lamp [1] energy per second transferred by light from halogen lamp is 25 J (= 25% of 100 W) compared to 1.8 J (= 90% of 2 W) for each LED [1] so 14 LEDs (~ 25/1.8) give same light output as one 100 W lamp [1]

3a $\frac{36\,\text{MJ}}{4 \times 3600\,\text{s}}$ [1] = 2.5 kW [1]

b $\frac{36\,\text{MJ}}{2000\,\text{W}}$ [1] = 18 000 s **or** 5 hours [1]

4a 1.5 A × 230 V [1] = 345 W [1]

b 345 W × (130 × 60 × 60) s [1] = 161 MJ [1]

P6.1

1a $0.024\,m^3$ ($= 0.80\,m \times 0.60\,m \times 0.05\,m$) [1]

b $\frac{60\,kg}{0.024\,m^3}$ [1] $= 2500\,kg/m^3$ [1]

2a $136\,g - 48\,g$ [1] $= 88\,g$ [1]

b $\frac{88\,kg}{80\,cm^3}$ [1] $= 1.1\,g/cm^3$ [1]

3a i $0.00040\,m^3$ ($= 0.10\,m \times 0.080\,m \times 0.05\,m$) [1]

ii $\frac{0.76\,kg}{0.00040\,m^3}$ [1] $= 19000\,kg/m^3$ [1]

b $19000 = \frac{0.0015}{V}$

$V = \frac{0.0015\,kg}{19000\,kg/m^3} = 7.9 \times 10^{-8}\,m^3$ [1]

thickness $t = \frac{7.9 \times 10^{-8}\,m^3}{0.15\,m \times 0.12\,m}$ [1] $= 4.4 \times 10^{-6}\,m$ **or** 0.0044 mm [1]

4 Use top pan balance to measure mass of bolt, [1] fill measuring cylinder half-full of water and measure volume of water in it, [1] tie bolt on thread and gently lower fully into water and mesure volume [1], volume of bolt is equal to the volume of water with bolt minus inital volume of water [1] use density = mass/volume to calculate density of bolt from its mass and volume [1]

P6.2

1a Freezing [1]

b Melting [1]

2a Condensation [1]

b Melting [1]

c Freezing [1]

3a Particles start to move about at random, [1] no longer in fixed positions [1]

b Particles in water vapour move at random and not in contact with each other except when they collide, [1] when water vapour condenses on a cold surface, vapour particles lose energy when they collide with surface and stay on surface as a film of liquid, [1] particles in film move at random in contact with each other [1]

4 Same mass of ice cube and water in beaker after ice cube has melted, [1] density = mass per unit volume so if volume of ice cube > volume of melted water, density of ice < density of water [1]

5 Particles in gas much more energetic and move faster and spaced further apart than particles in solid or liquid, [1] for given mass of gas, particles occupy much greater volume than equal mass of same substance in liquid or solid state, [1] density = mass/volume, [1] density of gas much less than density of same substance as liquid or solid [1]

P6.3

1 Any two from: boiling takes place at a certain temperature whereas evaporation occurs from a liquid at any temperature, [1] boiling takes place throughout liquid whereas evaporation from surface only, [1] evaporation can cause liquid to cool whereas boiling does not [1]

2a i Graph with suitable scales, [1] correctly plotted points, [1] best fit line with flat section from 150 s to about 240 s [1]

ii 79 °C [1]

b 60 °C: solid, [1] at 79 °C: begins to melt, [1] after 90 s all melted and liquid temperature then rises to above 90 °C [1]

3 Salt and water form solution which will not freeze unless temperature drops below freezing point of solution, [1] so no ice forms on road unless temperature drops below this freezing point, [1] grit provides friction between tyres and ice to help stop vehicles sliding [1]

4 Particles move randomly in contact with each other as temperature falls from 80 °C to 75 °C, [1] as temperature falls particles lose energy and move more slowly until at 75 °C they stop moving around and substance changes from liquid to solid, [1] at 75 °C particles become fixed in position and vibrate, [1] once all substance has changed state, temperature falls from 75 °C to 70 °C and vibrations of particles less vigorous [1]

P6.4

1a Liquid: particles close together and move about, not in fixed positions [1]

b Gas: particles far apart and move about [1]

c Solid: particles vibrate about fixed positions and close together [1]

d Does not exist: particles that vibrate about fixed positions are in solid and therefore not far apart [1]

2 When solid heated to its melting point particles gain KE and vibrate more about fixed positions, [1] at melting point, particles gain enough energy to break away from each other and move about, [1] particles that break free are in liquid state as they move about in contact with each other [1]

3 Internal energy transferred to solid at its melting point gives molecules enough energy to overcome strong forces of attraction holding them together in solid structure, [1] PE increases as particles break free from each other [1]

4 Heat energy transferred by heating from warm water to ice, [1] water cools and water particles move more slowly so they lose KE, [1] ice melts because it gains internal energy and particles in ice gain enough PE to break free from each other, [1] when ice melts, melted water from ice and warm water mix so particles from the ice, on average, gain KE and particles from warm water lose KE [1]

P6.5

1a $0.068\,kg - 0.024\,kg = 0.044\,kg$ [1]

b $15000 = 0.044 \times L$

$15000\,J/0.044\,kg)$ [1] $= 340\,kJ/kg$ [1]

2 $m_w = 0.152\,kg - 0.144\,kg = 0.008\,kg$ [1]

$L = 18400\,J/0.008\,kg$ [1] $= 2.3\,MJ/kg$ [1]

3a $E_1 = 0.120 \times 4200 \times (15 - 9)°C$ [1] $= 3024\,J$ [1]

b $E_2 = 0.008 \times 4200 \times 9$ [1] $= 302\,J$ [1]

c Energy transferred to melt ice $= 3024 - 302 = 2722\,J$ [1]

specific latent heat of fusion of water $= \frac{2722\,J}{0.008\,kg}$ [1]

$= 340\,kJ/kg$ [1]

4 $E = 0.100\,kg \times 2.25\,MJ/kg = 225000\,J$ [1] $t = \frac{225000\,J}{3000\,W}$ [1]

$= 75\,s$ [1]

P6.6

1a Increases [1]

b Unchanged [1]

c Increases [1]

2 Smoke particles move at random due to random impacts of air molecules, [1] if gas temperature increases, gas molecules move faster on average so impacts are harder and number of impacts per second increases, [1] so smoke particles move faster [1]

3 Gas pressure stops increasing (or decreases) when valve opens, [1] number of gas molecules in cylinder decreases, [1] so number of impacts they make per second on cylinder's internal surface decreases and gas pressure stops increasing (or decreases) [1]

4a Unless water is stirred, (hot water rises and so) temperature of water differs in beaker, [1] thermometer does not measure average temperature of water [1]

b Air in flask before sealed is at atmospheric pressure, [1] so pressure gauge reads atmospheric pressure before it is sealed [1]

P6.7

1 Volume increases and pressure decreases [1]

2a $\frac{100\,000\,\text{Pa} \times 0.000\,20\,\text{m}^3}{50\,000\,\text{Pa}}$ [1] $= 0.000\,40\,\text{m}^3$ [1]

b $\frac{100\,000\,\text{Pa} \times 0.000\,30\,\text{m}^3}{0.000\,15\,\text{m}^3}$ [1] $= 200\,000\,\text{Pa}$ [1]

c $\frac{100\,000\,\text{Pa} \times 0.000\,60\,\text{m}^3}{120\,000\,\text{Pa}}$ [1] $= 0.000\,50\,\text{m}^3$ [1]

d $\frac{60\,000\,\text{Pa} \times 0.000\,45\,\text{m}^3}{0.000\,15\,\text{m}^3}$ [1] $= 180\,000\,\text{Pa}$ [1]

3 initial volume $V_1 = 120\,\text{cm}^3$, final volume $V_2 = 100\,\text{cm}^3$ [1]
$\frac{100\,000\,\text{Pa} \times 120\,\text{cm}^3}{100\,\text{cm}^3}$ [1] $= 120\,\text{kPa}$ [1]

4 WD on air by applied force, [1] so internal energy of trapped air increases, so temperature of air increases, [1] as compression rapid, very little energy transfer to surroundings takes place [1]

P7.1

1a Beta is negatively charged and gamma is neutral [1] beta is a particle/electron and gamma is an electromagnetic wave [1]

b Alpha is made of two protons and two neutrons [1] neutrons have no charge and protons have a charge of +1 [1]

2a i Alpha [1]

ii Beta or gamma [1]

b Gamma [1]

3 Atoms have unstable nuclei, [1] these nuclei become stable by emitting radiation [1]

4a Substance emits (ionising) radiation [1] so radioactive [1]

b Paper stopped most radiation from substance reaching Geiger counter, [1] paper absorbed radiation, [1] so must be alpha radiation [1]

P7.2

1 Any three from: nucleus much smaller than atom, [1] nucleus positively charged, [1] mass of atom concentrated in nucleus [1] all positive charge of atom concentrated in nucleus [1]

2a $1 \times 10^{-10} \times \frac{1}{10\,000} = 1 \times 10^{-14}\,\text{m}$ [1]

b $10\,000 \times 0.033 = 330\,\text{m}$ [1]

3a **B** [1]

b **A**: attracted by nucleus [1] **C**: unaffected by nucleus [1] **D**: repelled in wrong direction by nucleus [1]

4a i Atoms not indivisible, [1] atoms contain negatively charged electrons [1]

ii Any two from: nuclear: all positive charge concentrated in nucleus much smaller than atom, plum pudding: positive charge spread out throughout atom, [1] nuclear: most mass concentrated in nucleus, plum pudding: mass spread out throughout atom [1] nuclear: most atom empty space, plum pudding: no empty space [1]

b Nuclear model explains why some alpha particles scattered through large angles, [1] in plum pudding model such large-angle scattering should not be observed [1]

5a Similarity: proton and neutron have about same mass (**or** both found in nucleus) [1] difference: proton is charged whereas neutron has no charge [1]

b He nucleus contains 4 particles (neutrons + protons) whereas H nucleus only contains one (a single proton), [1] 2 protons particles in He nucleus because He nucleus has twice as much charge as H nucleus, [1] therefore other 2 particles in He nucleus are neutrons [1]

P7.3

1a i 6p + 6n [1]

ii 27p + 33n [1]

iii 92p + 143n [1]

b 4p [1] 10n [1]

2a Any three from: same number of protons/6 protons [1] different number of neutrons/C14 has two more neutrons [1] nucleus has the same charge [1] nucleus of C14 has more mass [1]

b Mass stays the same, [1] charge increases [1] charge increases [1]

c ${}^{-6}_{14}\text{C} \rightarrow {}^{-7}_{14}\text{N} \rightarrow -{}^{-1}_{0}\text{e}$ [2]

3a ${}^{-92}_{238}\text{U} \rightarrow {}^{-90}_{234}\text{Th} + {}^{-2}_{4}\text{He}$ [2]

b ${}^{64}_{29}\text{Cu} \rightarrow {}^{64}_{30}\text{Zn} + {}^{0}_{-1}\text{e}$ [2]

4 ${}^{210}_{83}\text{Bi}$ [1] $\rightarrow {}^{210}_{84}\text{Po}$ [1] $+ {}^{0}_{-1}\text{e}$ [1]

P7.4

1a Stops irradiation of nearby people or objects [1]

b Alpha [1]

c Alpha, beta [1]

2a i Gamma [1]

ii Alpha [1]

iii Beta [1]

b i Gamma [1]

ii Alpha [1]

3a Can knock electrons from atoms, [1] this ionisation damages cells (**or** kills cell **or** affects genes in cell which can be passed on if cell generates more cells) [1]

b Alpha is highly ionising [1] but not penetrating, [1] outside the body it is low risk as cannot penetrate body [1] inside the body alpha radiation absorbed by cells [1] much greater risk of damage to cells [1]

4 Very little γ radiation absorbed by foil, it would all pass straight through [1] so thickness of foil would not affect detector reading [1]

P7.5

1a Average time for number of nuclei in sample of isotope to halve/time taken for activity/count rate to halve [1]

b 190 cpm [1]

2a 8 days [1]

b $n = \frac{24}{8} = 3$ [1] $800 \times \frac{1}{2^3}$ [1] $= 100$ cps [1]

3a i 4 milligrams (= 8 mg/2) [1]

ii 1 milligram (= 8 mg/2^3) [1]

b Fraction $= \frac{0.25}{8} = \frac{1}{32}$ [1] $n = 5$ [1] time $= 5 \times 15 = 75$ hours [1]

4a i 160 million atoms [1]

ii $\frac{1}{32}\left(=\frac{1}{2^5}\right)$ [1]

iii Number remaining = 320 million/2^5 [1] = 10 million atoms [1]

5 25% = $\frac{1}{4}$ [1] 2 half-lives [1] ∴ the wood is 11 200 years old (= 2 × 5600 yrs) [1]

P7.6

1a Beta or gamma, [1] can be detected outside body [1]

b Gamma, [1] radioactive source injected into patient to enter organ to be imaged so needs to do least damage whilst in body, [1] gamma radiation passes through body tissue and detected using gamma camera [1]

2 Any two from: food, drink, radon rocks/ground [1]

3a Small 'seeds' of radioactive isotope placed in tumour, [1] radiation from isotope destroys cancer cells, [1] half-life long enough to destroy cancer cells, but short enough so most of the unstable nuclei decayed soon afterwards [1]

b Beta **or** gamma [1]

4a Too short: radioactive isotope decays too much before scan completed, [1] too long: patient exposed to ionising radiation unnecessarily [1]

b Yes, this is important to make sure that the implant is not ionising [1] but also does not decay too soon [1]

5a Any three from: emits radiation detectable outside body (e.g., gamma), [1] non-toxic, [1] short half-life (1–24 hours), [1] decays into stable isotope [1]

b Stable isotope in body (or elsewhere) not dangerous [1] whereas unstable isotope harmful [1] as it emits ionising radiation harmful to body [1]

P7.7

1a Nucleus splits into two fragments, [1] releases energy [1] and 2/3 neutrons [1]

b Nucleus absorbs neutron without undergoing fission, [1] forms unstable nucleus which decays [1]

2 (in order) **B**, **A**, **C**, **D** [1]

3a To absorb fission neutron [1] and keep chain reaction under control by maintaining even rate of fission [1]

b More fission neutrons absorbed, so number of fission neutrons in reactor core decreases, [1] rate of release of energy due to fission decreases [1]

c Thick steel withstands very high temperature and pressure in core, [1] thick concrete walls absorb ionising radiation [1] that escapes through steel walls [1]

4a **A** and **D** [1]

b Undergone fission and released neutrons and energy [1]

c **C** and **E** [1]

P7.8

1a Formation of a heavier nucleus when two lighter nuclei fuse/join together [1]

b They repel [1]

2a High temperature [1] high pressure [1]

b Energy output is less than energy input so does not produce any energy overall [1]

3 Advantages: any two from: nuclear fusion fuel easily available [1] fusion products non-radioactive (**or** less radioactive than) fission products [1] fusion stops if plasma out of control [1]
Disadvantages: any two from: very large current needed to heat plasma to start fusion, [1] plasma difficult to control, [1] at present, fission reactors produce far more power than fusion reactors [1]

4a 1 p and 1 n [1]

b ${}^{2}_{1}\text{H} + {}^{1}_{1}\text{p} \rightarrow {}^{3}_{2}\text{He}$ [1]

c ${}^{3}_{2}\text{He} + {}^{3}_{2}\text{He} \rightarrow {}^{4}_{2}\text{He} + {}^{1}_{1}\text{p} + {}^{1}_{1}\text{p}$ [2]

P7.9

1a i Hazardous and therefore a danger to people and animals if it escapes [1]

ii Contains radioactive isotopes with long half-lives [1]

b Absorbed by surrounding tissues [1] and could damage or kill cells in body or cause cancer, [1] outside less dangerous as α radiation has no penetrating power [1]

2a May be more concentrated than outdoors and people could breathe it in, [1] lungs then exposed to α radiation, [1] ionising effect of α particles in tissue damages cells (**or** kills cells **or** causes cancer) [1]

b Install pipes under house and pump radon gas out of ground before it seeps into house, [1] top of outlet pipe from pump needs to be high up outside house [1]

3 Benefits: any two from: no greenhouse gas emissions, [1] reliable and secure electricity supplies, [1] large-scale generation from small sites compared with renewable supplies that take up much larger areas (**or** other valid points) [1] drawbacks: any two from: long-term storage of nuclear waste, [1] possible escape of radioactive substances into environment, [1] impracticality of fusion reactors, (**or** other valid points) [1]

4 Total = 2590 μSv [1]
Total from man-made source = (364 + 5 + 3 + 8) = 380 μSv [1] % = (380/2590) × 100 = 14.7 % [1] = 15 % to 2 s.f. [1]

P8.1

1a Size of quantity [1]

b Scalar has magnitude only, vector has direction too [1]

2 Speed, energy, mass [1]

3 Scale diagram where **A** is 1.25 times longer (15/12) than **B** e.g **A** is 12 mm and **B** is 15 mm [1] A pointing in opposite direction to **A** [1]

4a Depends on 48 N arrow length: e.g., 48 mm arrow, the scale is 10 mm = 10 N [1]

b Arrow **B** drawn to same scale as A e.g. using 10 mm = 10 N then arrow length is 36 mm [1] arrow pointing down and along dashed line [1]

P8.2

1a Decelerates [1]

b Force equal and opposite to force road exerts on each tyre [1]

2a 50 N upwards [1]

b 200 N [1]

3a Forces equal in magnitude to each other and opposite in direction [1] because book presses on table and table exerts equal and opposite force on book [1]

b Forces vertically downwards but force of table on floor greater than force of book on floor [1] because floor supports weight of table and book, table only supports book [1]

4a 500 N downwards [1]
b 500 N upwards [1]
c 500 N upwards [1]

P8.3

1 Glider makes contact with track and stops moving along it, [1] glider stops because friction opposes its motion [1]
2a Opposite in direction to velocity [1]
b 0< >N [1]
3a Force of mud on car greater than force on car from tractor [1]
b 300 N [1]
4 Car represented by a circle of square, [1] up arrow and down arrow equal in length, [1] up arrow labelled 'reaction force' and down arrow labelled 'weight', [1] arrow labelled 'friction' and arrow labelled 'engine force' in opposite directions, [1] friction arrow longer than engine arrow [1]

P8.4

1a i Increased [1]
ii Unchanged [1]
iii Reduced to a quarter [1]
b $M = 72 \times 0.25$ [1] 18[1] Nm [1]
2a Anticlockwise [1]
b i Increased [1]
ii Decreased [1]
3a Moment of applied force about pivot is greater the longer the handle, so greater turning force exerted on nail [1]
b Rust on hinge increases frictional forces in hinge, [1] so greater moment so more turning force must be applied to door to overcome moment of frictional forces at hinge [1]
4 $18\,\text{Nm} = F \times 0.25$ [1]
$F = \frac{18}{0.25}$ [1]
72 N [1]

P8.5

1 Force applied to each handle of T bar acts at much greater distance from axis of T bar than force of nut on T bar, so force applied to nut much greater than force applied to T bar, [1] as moment of forces applied to T bar must least equal moment of nut on T bar to turn nut [1]
2 $4.8 = F \times 0.04$
$F = 4.8/0.04$
120 N [1]
3 Uphill, climb force of wheel on road greater than on flat road, [1] turning effect of chain on gear wheel opposed by turning effect of frictional force of road on wheel, [1] if use larger gear wheel, turning effect of force of chain on gear wheel increases so force of wheel on road increases [1]

P8.6

1a i $W_1 \times 0.30 = 6.0 \times 0.15$ [1]
$W_1 = \frac{6.0 \times 0.15}{0.30} = 3.0\,\text{N}$ [1]
ii $4.0 \times 0.6 = 12.0 \times d_2$ [1]
$d_2 = \frac{4.0 \times 0.6}{12} = 0.2\,\text{m}$ [1]
b In both examples line of action of effort at greater perpendicular distance from pivot than corresponding distance for load, [1] smaller effort therefore gives equal and opposite moment about pivot to load's moment [1]
2a $2.0 \times 0.24 = W \times 0.08$ [1]
$W = \frac{2.0 \times 0.24}{0.08}$ [1]
$W = 6.0\,\text{N}$ [1]
b Check results are repeatable, [1] identify any anomalous results and discard when calculating the mean [1]
3a $(0.4 \times 2) + (0.6 \times 1)$ [1] $= 1.4\,\text{Nm}$ [1]
b $5 \times d = 1.4$ [1]
$c = \frac{1.4}{5} = 0.28\,\text{m}$ [1]
4 $d_0 \times 120 = 6d_0 \times F$
$120 = 6F$
$F = 20\,\text{N}$ [1]

P8.7

1a Both forces act in the same direction (on the same object) [1]
b Both forces act in opposite direction (on the same object) [1]
2 Figure 6: 50 N vertically upwards [1]
Figure 7: 500 N up the slope [1]
3a Forces drawn and labelled to scale [1] resultant drawn and labelled [1] 5.0 N at 37° to 4.0 N force [2]
b Forces drawn and labelled to scale [1] resultant drawn and labelled [1] 6.1 N at 26° to 4.0 N force [2]
c Forces drawn and labelled to scale [1] resultant drawn and labelled [1] 6.5 N at 28° to 4.0 N force [2]
4 Forces drawn and labelled to scale [1] resultant drawn and labelled [1] 5400 N (to 2 s.f.) [1]

P8.8

1 10 N force drawn to scale at correct angle [1] horizontal and vertical components drawn [1] lengths measured [1] values of 7.1 N for both [1]
2 800 N force drawn to scale at correct angle [1] horizontal and vertical components drawn [1] horizontal component = 690 N to 2 s.f [1]
3 Not enough friction on ladder at floor, [1] so weight pulls ladder down [1]
4a Correct diagram [1] with parallel component = 25 N [1], perpendicular component = 43 N [1]
b Friction on box equal and opposite to parallel component of weight (25 N) [1]

P9.1

1a i Does not change [1]
ii Constant gradient [1]
b i $\frac{30\,000\,\text{m}}{1000\,\text{s}}$ [1] = 30 m/s [1]
ii 500 s [1]
iii $\frac{20\,000\,\text{m}}{1500\,\text{s}}$ [1] = 13.3 m/s [1]
2a $\frac{1800\,\text{m}}{60\,\text{s}}$ [1] = 30 m/s [1]
b 30 m/s × 3000 s [1] = 9000 m [1]
c $\frac{3300\,\text{m}}{30\,\text{ms}}$ [1] = 110 s [1]
3 $s = 21 \times 6 \times 60$ [1] = 7560 m [1]
$v = \frac{7560}{(7 \times 60)} = 18\,\text{m/s}$ [1]
4a $\frac{360\,000\,\text{m}}{160 \times 60\,\text{s}}$ [1] = 37.5 m/s [1]
b $\frac{180\,000\,\text{m}}{40\,\text{m/s}}$ [1] = 4500 s = 75 minutes [1]

P9.2

1a Speed is distance travelled ÷ time taken regardless of direction, velocity is speed in a given direction [1]

b Distance apart = (30 m/s − 20 m/s) × 300 s [1] = 2400 m [1]

2 $\frac{28\,\text{m/s} - 8\,\text{m/s}}{16\,\text{s}}$ [1] = 1.25 m/s^2 [1]

3a i As it left motorway [1]

ii When travelling at constant velocity [1]

b $a = \frac{v-u}{t}$ gives $2.0 = \frac{v-7}{10}$ [1] $v - 7 = 2.0 \times 10 = 20$ [1]

∴ $v = 27$ m/s [1]

4a $\frac{9.2\,\text{m/s} - 0}{3.1\,\text{s}}$ [1] = 2.97 m/s^2 [1]

b $\frac{100\,\text{m}}{10.4\,\text{s}}$ [1] = 9.6 m/s [1]

P9.3

1 i **B**

ii **A**

iii **D**

iv **C** (all correct = [2] any 3 correct = 1])

2a Acceleration = $\frac{8}{20}$ [1] = 0.4 m/s^2 [1]

b Acceleration = $\frac{4}{20}$ [1] = 0.2 [1] = −0.2 m/s^2 [1]

3a i **A** [1]

ii **C** [1]

b **B** [1]

4a Area **C** = $\frac{1}{2}$ × 4 m/s × 20 s [1] = 40 m [1]

b Area **A** = 8 × 20 = 160 m

area **D** = $\frac{1}{2}$ × 6 m/s × 20 s [1] = 60 m [1]

difference = 160 m − 60 m = 100 m [1]

P9.4

1 Increases gradually [1] from 0 at start [1]

2a Constant acceleration from rest to 8 m/s for 40 s [1] then constant deceleration for last 20 s [1]

b i $a = \frac{8\,\text{m/s} - 0\,\text{m/s}}{40\,\text{s}}$ [1] = 0.20 m/s^2 [1]

ii $a = \frac{0\,\text{m/s} - 8\,\text{m/s}}{20\,\text{s}}$ [1] = −0.40 m/s^2 [1]

c Total distance = $\left(\frac{1}{2} \times 40 \times 8\right) + \left(\frac{1}{2} \times 20 \times 8\right)$ [2]

= 160 + 80 = 240 m [1]

3a Suitable scales [1] correctly plotted [1] best fit line drawn [1]

b $\frac{40\,\text{m/s} - 0\,\text{m/s}}{20\,\text{s}}$ [1] = 2.0 m/s^2 [1]

c i $\frac{1}{2}$ × 40 m/s × 20 s [1] = 400 m [1]

ii 40 m/s × 10 s [1] = 400 m [1]

4 $v^2 - 0 = 2 \times 2 \times 144$ [1]

$v^2 = 576$ [1]

$v = 24$ m/s [1]

P10.1

1a The acceleration of an object is proportional to the resultant force acting on it [1] and inversely proportional to the mass [1]

b $F = 80 \times 8$ [1] = 640 N [1]

2a 16 N [1]

b 40 kg [1]

c 12 m/s^2 [1]

3a 1500 kg × 2 m/s^2 = 3000 N [1]

b i $F = 300 \times 2 = 600$ N [1]

ii 3000 N − 600 N [1] = 2400 N [1]

4a Total mass greater in the second case [1] and force same so acceleration is less [1]

b $F = 0.60m = 0.48(m + 0.5)$ [1] gives $m = 2.0$ kg [1]

P10.2

1a Initial resultant force is equal to weight [1]

b Frictional force is less than weight [1]

c Zero [1]

d Zero [1]

2a W = 80 × 9.8 [1]

W = 784 N [1]

b W = 80 × 1.6 [1]

W = 128 N [1]

c 2058 = m × 9.8 [1]

m = 2058/9.8 210 kg [1]

Moon W = 210 × 1.6 [1]

W = 336 N [1]

3a Resultant force = weight – frictional force [1] frictional force due to parachute increases with speed [1] so resultant force on parachutist decreases, [1] when frictional force = weight, resultant force = 0 and parachutist moves at terminal velocity [1]

b i W = (70 + 20) × 9.8 = 882 N [1]

ii 882 N upwards [1]

4a Gradient measured at 0.10 s [1] = 5.2 m/s^2 [1]

b At 0.1 s resultant force = m × 5.2 [1]

resultant force = weight − drag force [1]

weight = mg = 9.8 m [1]

5.2 m = 9.8 m − drag

drag = 4.6 m [1]

drag/weight = 4.6 m/9.8 m = 0.47 [1]

P10.3

1a Braking distance [1]

b Thinking distance [1]

c Braking distance [1]

2a i (21 − 15) = 6.0 m [1]

ii (37.5 − 13.5) = 24.0 m [1]

iii 30.0 m [1]

b (30 m/s × 0.8 s) − (15 m/s × 0.8 s) [1] = 12 m [1]

3a Thinking distance is proportional to speed [1] as reaction time is constant [1]

b Braking distance divided by v^2 same for all three speeds [1] so braking distance proportional to v^2 [1] so claim is valid [1]

4a $31^2 - 0 = 2 \times a \times 75$ [1]

961 = 150a

a = 961/150 [1]

a = 6.4 m/s^2

b 1500 kg × 6.4 m/s^2 [1] = 9600 N [1]

P10.4

1 Momentum = mass × velocity, kg m/s [1]

2a 20 kg × 20 m/s = 400 kg m/s [1]

b 400 = 800 × v

$v = \frac{400}{800}$ = 0.5 m/s [1]

3a Equal and opposite forces [1]

b Equal and opposite momentum [1]

c v 80 kg skater = three-quarters v 60 kg skater [1] in opposite direction [1]

d Total momentum = 0 [1]

4a 60 kg × 2 m/s = 120 kg m/s [1]

b $80v$ = 120 kg m/s [1] ∴ v = 120/80 [1] = 1.5 m/s [1]

P10.5

1a 1000 kg × 5.0 m/s = 5000 kg m/s [1]

b (1000 kg × v) + (1500 kg × v) = 2500v [1] ∴ 2500v = 5000 [1] ∴ v = 2.0 m/s [1]

2 (70 kg × 5 m/s) + 0 = 350 kg m/s [1]
(70 + 1.2) × v = 350 kg m/s [1]
$v = \frac{350}{71.2} = 4.9$ m/s [1]

3a −12v = 600 × 0.5 kg m/s [1] ∴ v = −600 × 0.5/12 [1] = −25 m/s [1]

b Less [1]

4 Momentum before impact = (0.80 kg + m) × 1.1 m/s = (0.88 + 1.1 m) kg m/s [1] momentum after impact = (m + 0.80) kg × 0.70 m/s + (0.80 kg × 0.70 m/s) [1]
= 0.70m + 1.12 [1]
∴ 0.70m + 1.12 = 0.88 + 1.1m [1] ∴ 0.40m = 0.24 [1]
∴ m = 0.24/0.40 = 0.60 kg = [1]

P10.6

1 Seat belt increases time taken to stop person [1] so change of momentum per second less so force on person less [1]

2 Change in velocity = 0 − 18 = −18 m/s [1]
$F = \frac{0.12\text{ kg} \times -18\text{ m/s}}{0.0003\text{ s}}$ [1]
F = −7200 N [1]

3a i F = change in momentum/time [1]
$F = \frac{800\text{kg} \times 30}{6} = 4000$ N [1]

ii F = change in momentum/time [1]
$F = \frac{800\text{kg} \times 30}{30} = 800$ N [1]

b Force = change of momentum/time taken or rate of change of momentum [1] change of momentum same but time taken much less, so force is much greater [1]

4 Impact time on cushioned surface longer than on hard floor, [1] a given change of momentum/change of momentum per second therefore less in fall on cushioned floor than on hard floor so impact force less [1]

5a Initial momentum = 2000 kg × 12 m/s = 24 000 kg m/s [1] = final momentum, [1] = total mass × the velocity after impact, ∴ = 24 000 kg m/s ÷ 12 000 kg = 2 m/s [1]

b i Deceleration = $\frac{\text{change of velocity}}{\text{time taken}}$
$= \frac{2\text{ m/s} - 12\text{ m/s}}{0.3\text{ s}}$ [1]
= (−) −33 m/s² [1]

ii Force = $\frac{\text{change of momentum}}{\text{time taken}}$
$= \frac{-20\,000\text{ kg m/s}}{0.3\text{ s}}$ = [1]
= (−) 67 000 N [1]

P10.7

1 Protects cyclist's head in collision (**or** if cyclist falls off cycle and head hits ground) because when impact occurs helmet increases time taken to decelerate head, [1] so it reduces change of momentum per second [1] and reduces impact force [1]

2 If car suddenly stopped, child would press against back of car seat spreading out force [1] and back of car seat would prevent child from being thrown forwards [1]

3 It increases time taken to stop reducing force of impact because change of momentum takes longer, [1] force spread out across chest so effect less [1]

4 Increases impact time so reduces impact force [1] spreads force over a larger area to reduce impact [1] stretches in a collision to increase impact time further [1]

5a (2150 kg + 750 kg) × 9 m/s = 26 100 kg m/s [1]

b Momentum before impact = 750 kg × v [1]
∴ 750v = 26 100 [1] which gives $v = \frac{26\,100}{750} = 35$ m/s [1]

c Yes [1]

P10.8

1 Extension of spring directly proportional to force applied as long as limit of proportionality not exceeded [1]

2a Limit beyond which force no longer proportional to extension [1]

b Increase in length from its original unstretched length [1]

3a Does not return to original length when released [1]

b Rubber band returns to original length when released whereas polythene strip does not [1]

4a 80 mm [1]

b Spring constant equals the gradient of the graph [1]
Choose a two pairs of values e.g. (0,0) and (80,4) [1]
spring constant = $\frac{4 - 0}{0.08 - 0}$ = 50 N/m [1]

5a $\frac{5.0\text{ N}}{25\text{ N/m}}$ [1] = 0.20 m [1]

b Work done = elastic potential energy stored [1]
$= \frac{1}{2} \times 5 \times 0.22$ [1] = 0.1 J [1]

P11.1

1a Weight unchanged, area of hands less than area of feet [1] so pressure on hands during handstand more than pressure on feet when standing [1]

b Contact area of snowshoes much greater than contact area of soles of feet, [1] pressure = weight/contact area, so wearing snow shoes results in less pressure on snow [1]

2a $\frac{1200\text{ N}}{(0.60\text{ m} \times 0.40\text{ m})}$ [1] = 5000 Pa [1]

b $\frac{1200\text{ N}}{(0.40\text{ m} \times 0.05\text{ m})}$ [1] = 60 000 Pa [1]

3 45 000 Pa × 0.0002 m² = 9.0 N [1]
Therefore force must be greater than 9.0 N [1]

4 $\frac{1.8 \times 10^6}{18}$ = 100 000 N [1]
area in contact = $\frac{100\,000}{1.4 \times 10^6}$ [1]
= 0.07 m² [1]

P11.2

1a Pressure of water greater at bottom of dam than at top, [1] so dam thicker at bottom to withstand greater force due to water pressure there [1]

b Vertical distance from water tank to ground floor tap greater than distance to tap on higher floor, [1] water pressure increases with vertical distance below tank so water pressure at ground floor tap higher than pressure at higher tap [1]

2 Increase in depth = 1.2 − 0.4 = 0.8 [1]
increase in $p = 0.8 \times 1000 \times 9.8$ [1]
$p = 7840$ Pa [1]

3a 0.090 m × 1000 kg/m³ × 9.810 N/kg [1] = 882 900 Pa [1]

b 900 Pa × 0.0006 m² = 0.54 N [1]

4 Measure volume V of tap water, measure height H of water column from same level as bottom of oil column, [1] empty U-tube and repeat test using same volumes of sea water and oil as before, [1] density of sea water/density of tap water = H_{tap}/H_{sea} [1]

P11.3

1 Thin layer of gases, [1] same density [1]

2 Atmospheric pressure caused by air molecules colliding with each other and with surfaces, [1] each impact causes tiny force on surface, [1] many molecules hit surface each second so cause measurable pressure on surface, [1] decreases with height because number of molecules per unit volume decreases [1] (**or** pressure at any given altitude due to weight of air above [1] which decreases with increasing altitude so pressure decreases with increasing altitude [1])

3 100 000 Pa × 1.2 × 10^{-3} m² [1] = 120 N [1]

4 Change in height = 30 km − 10 km = 20 km [1]
Change in pressure = 25 kPa − 2 kPa = 23 kPa [1]
23 000 = 20 000 × 9.8 × density [1]
density = $\frac{23\,000}{20\,000 \times 9.8}$ = 0.12 kg/m³ [1]

P11.4

1 Ice less dense than water [1]

2a Pressure of water on lower half of ball greater than on top half so ball experiences upthrust, [1] volume of liquid displaced by ball is equal to volume of ball, [1] as water is much denser than air, weight of liquid displaced much greater than weight of ball [1] so upthrust on ball much higher than weight [1]

b Cork doesn't absorb water and much less dense than water, [1] upthrust on cork doesn't need to be very large to support its weight [1] so cork doesn't need to be fully immersed [1]

3a Upthrust acts on object in water, [1] reading on newton-meter less when in air as upthrust helps support it in water [1]

b Reading changes gradually from 5.2 N in air [1] to 4.7 N when all in water [1]

4a **A** has greatest density because it sinks and **B** and **C** both float, [1] **A** more dense than water, **B** and **C** less dense than water [1]

b **B** floats higher in water than **C** so difference between B's density and water is greater than **C**, [1] density of **B** less than density of **C** so lowest density is **B** [1]

5a Extra weight causes tube to float lower in water, [1] so length of tube above water decreases (linearly) as total weight is increased [1]

b Remove tube from water, add metal object (e.g., steel nail) of known weight to tube, [1] use ruler to measure length of tube (and cork) above water, [1] repeat five more times, each time adding another object of known weight, [1] record results in table, plot graph of L against W, [1] if prediction correct, graph has negative gradient [1] and straight line [1]

P12.1

1a Oscillations perpendicular to direction of energy transfer in transverse wave but parallel in a longitudinal wave [1]

b i Electromagnetic wave **or** waves on stretched string/wire water waves [1]

ii Sound waves [1]

c Particles displaced so closer together [1]

2a Transverse [1]

b i Along rope from one end to the other [1]

ii Oscillates in a direction perpendicular to energy transfer [1]

3 Stretch slinky out, move one end at right angles to slinky for transverse waves [1] or parallel to slinky for longitudinal waves [1]

4 Ball moves up and down repeatedly on surface [1] because waves consist of successive crests and troughs moving across surface, [1] each crest pushes ball up and each trough allows it to move down [1]

P12.2

1 The number crests passing a point in one second **or** the number of cycles of waves that pass a point in one second [1]

2a i Horizontally from **P** to next crest [1]

ii Vertically from **P** to midpoint [1]

b **P** moves vertically [1] down to a minimum then back to point **P** [1]

3a 2.0 Hz × 3.0 m [1] = 6.0 m/s [1]

b i $\frac{6.0\text{ m/s}}{1.0\text{ Hz}}$ [1] = 6.0 m [1]

ii 6.0 m/s × 60 s [1] = 360 m [1]

4a 340 m/s × 5.0 s [1] = 1700 m [1]

b $\frac{340\text{ m/s}}{3000\text{ Hz}}$ [1] = 0.11 m to 2 s.f. [1]

c $\frac{1}{3000}$ s [2]

P12.3

1 Incident angles are equal to reflected angles [1]

2 See P12.3 Figures 2 and 3: correct refraction, [1] correct directions, [1] refracted wavelength less than incident wavelength [1]

3a Velocity/speed of waves is higher in deeper water, [1] edge of wavefront in the deeper water speeds up, [1] part of wavefront in shallower water still moves at slower speed which leads to a change in the direction of the wavefront [1]

b All parts of the wavefront enter the deeper water at the same time, [1] all speed up at the same time [1]

4 Any four from: place sunglasses on white paper at fixed distance from light source in darkened room directly facing light source, [1] place light meter behind each lens in turn with meter as close as possible to lens without touching it, [1] mark position of meter at each lens on the paper, record each meter reading, [1] remove sunglasses without moving paper and record reading when meter at each marked position for each lens, [1] second reading – first reading for each lens shows effect of lens on light: the bigger the difference the less light transmitted [1]

P12.4

1 340 m/s × 4.0 s [1] = 1360 m [1]

2 Time delay = 0.24 s (= 2.4 s /10) [1] distance to wall and back = speed × time = 340 m/s × 0.24 s = 82 m, [1] s = 0.5 × 82 m = 41 m [1]

3a i Wavelength = $\frac{2}{3}$ × 1.24 [1]
= 0.83 m [1]

ii v = 180Hz wavelength = 0.83 m [1]
= 150 m/s to 2 s.f. [1]

b As the string moves up and down/vibrates [1] it pushes and pulls repeatedly on the air around it [1] this sends out vibrations through the air in waves of compressions and rarefactions [1]

c Any three from: particles in solid close packed [1] vibrations transferred more easily [1] energy transferred much faster [1] waves travel much faster in solid [1]

P12.5

1 Range of human hearing is from 20 to 20 000 Hz, [1]
The elephants produce sound of higher/lower frequencies than this [1]

2 Any two from: sound waves make the ear drum vibrate [1] vibrations passed through other parts/little bones/fluid in ear [1] detected by little hairs which send signal to the brain [1]

3 Distance to sea bed and back = 1500 m/s × 0.36 s = 540 m [1]
distance to sea bed = 0.5 × 486 m = 270 m [1]

4a Vibrating tuning fork makes table surface vibrate, [1] vibrating surface creates sound waves in much greater volume of air than tips of vibrating tuning fork [1]

b Suitable method e.g. switch on radio (or TV) and adjust sound to low level, go to adjacent room, close doors, [1] put ear to wall between the rooms and listen for radio, [1] sound waves from radio cause vibrations to pass through wall and send sound waves into ear [1]

P12.6

1 Ultrasound [1]

2a Organs have different density to surrounding tissue, [1] so ultrasound reflected at tissue/organ boundaries [1]

b Ultrasound is not ionising radiation but X-rays are, [1] X-rays harmful to living tissue, [1] ultrasound reflected at boundaries between tissues whereas X-rays are not [1]

c Some energy absorbed by body tissue as waves pass through body, [1] so amplitude of waves becomes smaller [1]

3 Distance = 1500 m/s × 0.40 s = 600 m [1]
so depth = 0.5 × 600 m = 300 m [1]

4a Time delay = $8 \times 4 \times 10^{-6}$ [1]
distance = $\frac{1}{2} \times 1500 \times 3.2 \times 10^{-5}$ [1]
distance = 0.024 m [1]

P12.7

1a Secondary (S), primary (P) [1]

b i P-wave [1]
ii S-wave [1]

c Wave speed increases gradually with depth, [1] so waves refracted, gradually changing direction, [1] bending towards surface because parts of each wavefront further from surface travel faster, [1] so wavefronts gradually move back towards surface [1]

2a See P12.7 Figure 1: inner core, [1] outer core, [1] mantle, [1] crust [1] (correct and labelled)

b i They are transverse waves so cannot travel through liquid outer core [1]
ii Speed changes at boundary, [1] because mantle is solid and outer core is liquid and seismic waves travel faster through solids than through liquids [1]

3a $\frac{5000\text{ km}}{460\text{ s}} = 10.9\text{ km/s}$ [1]

b $\frac{5000\text{ km}}{460\text{ s} + 180\text{ s}} = 7.8\text{ km/s}$
∴ S-waves ≈ 3 km/s slower [1]

4 Any six from: there is a shadow zone where no seismic waves are detected, indicating Earth has liquid core, [1] S-waves cannot travel through liquid outer core, [1] P-waves travel slower in outer core than in mantle indicating outer core is liquid whereas mantle is solid, [1] refraction at mantle/core boundary, and direction change away from boundary and from surface indicates change of state, [1] after travelling through outer core waves refract towards boundary when they travel into mantle, [1] overall effect is they refract more than waves that only just miss outer core, defining nearest part of shadow zone, [1] furthest part of shadow zone defined by waves entering core and refracted less than other waves that enter core, [1] existence of solid inner core deduced as very weak P-waves reach shadow zone [1] after entering outer core and then reaching solid inner core where refracted into shadow zone [1]

P13.1

1a Radio waves [1]
b Same for all electromagnetic waves [1]
c X-rays [1]
d Microwaves [1]

2a Radio, infrared, X-rays, gamma rays [1]
b microwaves, visible light, ultraviolet [1]

3a $\frac{300\,000\,000\text{ m/s}}{600\,000\,000\text{ Hz}}$ [1] = 0.50 m [1]

b $\frac{300\,000\,000\text{ m/s}}{0.30\text{ m}}$ [1] = 1000 MHz [1]

4 All electromagnetic waves travel at same speed in space, [1] gamma rays and visible light travel same distance and emitted at same time so reach Earth at same time [1]

P13.2

1a i Radio waves [1]
ii Light [1]
b i Microwaves [1]
ii Radio waves [1]

2a Handset signals would interfere with mobile phone signals so calls less clear [1]
b Other signals might 'mask' emergency services signals [1] making vital conversations difficult to listen to [1]

3 $\frac{300\,000\,000\text{ m/s}}{2400\,000\,000\text{ Hz}}$ [1] = 0.125 m [1]

4a Reflects microwaves from transmitter into receiver [1]
b i Place receiver directly in front of transmitter and measure signal detected, [1] place metal plate between receiver and transmitter and record meter reading [1]
ii Replace metal plate by cardboard and record reading again, [1] if reading not zero, microwaves can pass through [1]

P13.3

1a Visible light, infrared [1]
b Signals totally internally reflected so cannot escape from fibre except at receiver end, [1] radio signals travel through air so detected by any radio detector in their path [1]

2a Child's skull is thinner than adult [1] so more radiation passes through more easily (and causes a greater heating effect) [1]
b There is not enough evidence to certify the risks involved around using mobile phone [1]

3 Alternating current induced/electrons oscillate in the aerial [1] with same frequency as radio wave [1]

4a Wavelength = $\frac{300\,000\,000\,\text{m/s}}{105\,000\,000\,\text{Hz}}$ [1] = 2.8 m [1]

b Wavelength = $\frac{300\,000\,000}{1\,500\,000\,000}$ [1] = 0.2 m [1]
This is much smaller than wavelength/14 times smaller [1]

P13.4

1a X-rays and gamma rays (also radio waves) [1]
b Lead [1]
2a X-rays pass through crack but not through surrounding metal, [1] X-rays passing through crack darken photographic film, [1] crack appears as break in metal object shadow [1]
b Yes [1]
c Metal case stops low-energy X-rays but plastic does not [1] plastic allows X-rays to reach film inside case, giving more realistic account of exposure [1]
3a They can penetrate skin and cause skin cancer [1] and damage eyes/vision [1]
b Ultraviolet radiation from Sun causes skin damage and skin cancer, [1] suncreams absorb UV [1] to stop UV reaching skin [1]
4a Ionisation is process of making uncharged atoms become ions which are charged atoms [1] occurs when X-rays or gamma radiation pass through substances, [1] as X-rays and gamma radiation knock electrons out of uncharged atoms [1]
b i X-rays and gamma rays [1]
ii Ultraviolet radiation, X-rays and gamma rays [1]

P13.5

1a Absorbs X-rays, [1] otherwise X-rays pass through stomach so no details of stomach seen on X-ray film [1]
b Used to destroy cancerous tissue [1]
2a i Dense materials such as bone absorb X-rays and stop them reaching film, [1] X-rays that do not pass through dense material darken film, [1] when film developed, clear images of bones and other absorbing materials seen as X-rays did not reach these areas [1]
ii Shielding prevents X-rays reaching [1] and damaging cells [1] in parts of patient not under investigation [1]
b Ionisation can damage or kill cells or cause cell mutation and cancerous growth [1]
3a X-rays used for therapy have much shorter wavelengths/ greater energy [1]
b Not enough energy to destroy cancerous tumours [1]
4a A measure of the damage caused by ionising radiation to a person [1]
b About 0.3 millisieverts (≈ 13% of 2 mSv) [1]

P13.6

1 Electromagnetic radiation emitted from surface of objects due to their temperature (**or** electromagnetic radiation with wavelengths in range from about 700 nm to about 1 mm) [1]
2a City hotter than rural area [1] possible more energy dissipated in urban areas [1]
b Put hand near iron and see if it gets warm [1]
3a Temperature increased as heated so emitted more infrared radiation, increase in intensity of radiation emitted greater at shorter wavelengths than at longer wavelengths, [1] wavelength of light decreases from red to orange to yellow (to green and blue) so at higher temperature radiation emitted was orange-red rather than dull-red [1]
b i **Z** hottest, [1] because more light at shorter wavelengths [1]
ii **Y** coolest [1] because **X** emits orange and yellow which have shorter wavelengths than **Y** which emits red [1]

P13.7

1a Similarity: radiation emitted and absorbed is infrared, [1] difference: wavelengths of radiation emitted from Earth longer than wavelengths absorbed from the Sun [1]
b Radiation covers continuous range of wavelengths including longer wavelengths that are absorbed by greenhouse gases in atmosphere, [1] shorter wavelengths not absorbed and on a clear night they pass through atmosphere into space [1]
2a Any three from: infrared radiation from Sun passes through car windows to heat surfaces inside car, [1] hot surfaces heat air inside car so temperature in car rises, [1] the hot surfaces emit radiation with longer wavelengths which cannot pass through the glass, [1] rate of absorption greater than rate of emission so temperature rises
b Any two from: less infrared radiation from Sun reaches car so less infrared absorbed by surfaces inside in car, [1] so surfaces not as hot as inside car not as hot [1] the difference between rate of absorption and rate of emission not as great [1]
3 Infrared radiation emitted by Earth's surface has continuous range of wavelengths, [1] molecules of greenhouse gases in atmosphere absorb longer wavelength infrared from Earth and emit it back to surface, [1] making Earth warmer than if no greenhouse gases [1]

P14.1

1a i 20° [1] **ii** 40° [1]
b 42° [1]
2a A random error (not human) [1] any sensible suggestion e.g. mirror moved/ray box moved/reading protractor from an angle and not directly above [1]
b Mean = $\left(\frac{35 + 42 + 38 + 45}{4}\right)$ = 40° [1]
uncertainty = ±5° [1]
c Mean values approximately/equal to angle of incidence [1]
example of calculation e.g. $\left(\frac{19 + 18 + 20 + 22}{4}\right)$ = 19.8 [1]
could have a greater range of angles of incidence [1]
3a Specular: reflection of light from a smooth surface [1] so all rays reflect in same direction [1] diffuse: reflection of light from rough surface [1] scattered in all directions [1]
b Real image can be focused on screen, [1] virtual image can't [1]
4a 1st reflected ray correct [1] 2nd reflected ray correct relative to 1st [1]
b 180° [1]
c As two mirrors are perpendicular to each other, angle of incidence for second reflection = 90° – first angle of incidence, [1] adding both angles of incidence = 90°, [1] adding both reflected rays = 90° so total = 180° [1]

P14.2

1a Decrease [1]
b Zero [1]
c Smaller [1]
2a First refractions towards normal and second away from normal, [1] rays before and after passing through block parallel to each other [1]
b First refraction towards normal, [1] 2nd refraction away from normal [1]

3a Both refractions drawn correctly, [1] angles correct to within 2° [1]

b As rays from bottom of pool refract away from normal at the surface, [1] they appear to travel straight from a point above the bottom [1]

4 All colours of visible spectrum from red to violet seen on screen, [1] violet furthest from original direction of beam [1] because prism refracts each light colour a different angle, refracting violet light most and red light least [1]

P14.3

1 White page reflects all wavelengths of light [1] black ink absorbs all wavelengths of light [1]

2 Cover appears black because blue absorbs all incident red light and reflects none, [1] title appears red because white reflects red [1]

3a Dress appears green as reflects green [1] hat appears green because silver reflects all light [1]

b Silver reflects blue light so hat appears blue, dress appears black as absorbs blue wavelengths [1]

4 Any four from: place meter facing lamp emitting white light in dark room and record reading, [1] place block so light passes at 90° through it to meter and record reading, [1] calculate ratio of these two readings, [1] rotate block 90° so light travels through a different thickness and record meter reading, [1] remove block and check first reading unchanged then calculate ratio of third reading to first reading, [1] if the two ratios differ it is because amount of absorption depends on thickness of block, [1] lower ratio corresponds to thicker block [1] as absorption greater [1]

P14.4

1 A convex lens is thicker in the middle and brings light to a focus [1] a concave lens is thinner in the middle and spreads out the light [1]

2a Upright [1], magnified, [1] and virtual [1]

b Inverted, magnified, and real [1]

c Attempt at measurement e.g. flower diameter approximately 3 mm, in magnifying glass approx 6 mm [1]

magnification $= \frac{6}{3}$ [1] $\approx \times 2$ [1]

3a Attempt at measurement e.g. length of bird tip to tail 4 mm in image 11 mm [1]

Magnification $= \frac{11}{4} = 2.8$ [1] $\approx \times 3$ [1]

b Smaller [1] still inverted [1]

4 Any six from: use millimetre ruler to measure diameter of object and record measurement, [1] place metre ruler alongside lens holder and screen with zero end at front of lampbox, [1] adjust lens and screen so clear image of crosswires formed on screen, [1] using set square and metre ruler, [1] measure object distance and image distance [1] use millimetre ruler to measure diameter of image and calculate magnification, *M*, [1] enter measurements in table, adjust positions of lens and screen and repeat for five more image distances, [1] plot graph of *M* against image distance [1]

P14.5

1a Each construction rays drawn correctly, [3] the image drawn correctly in the correct position [1]

b i Real [1]

ii Diminished [1]

iii Inverted [1]

2a Three construction rays drawn correctly, image correct (as shown in Figure 2) [3]

b i Virtual [1]

ii Magnified [1]

iii Upright [1]

c Image always smaller than the object [1]

3a i Scale diagram [1] at least 2 of the 3 construction rays correctly drawn, [1] object and image correctly located [1]

ii Inverted and real [1] at 2F on other side of the lens, [1] image = object height so magnification = 1 [1]

b An inverter lens to invert image of distant object formed on screen by convex lens [1]

4a Diagram should show image and object correctly with image twice the height of object, [1] at least 2 construction rays drawn correctly [1]

b 1.8 cm [1]

c The nearer the object is to the focal point, the further the image from the lens [1] and the larger it would be, [1] no image formed at the focal point [1]

P15.1

1a N [1]

b N, [1] repels **X** because it has same polarity [1]

2a i N [1]

ii S [1]

iii unmagnetised [1]

b Pole of compass nearest tip of nail induces magnetism in nail with opposite pole at tip [1] so tip of nail always attracts end of plotting compass nearest it because they have unlike poles [1]

3a **X** = N, **Y** = S [1]

b Needle would turn (anticlockwise) as N-pole end of compass follows **Y** [1] until **X** attracts the S-pole more than **Y** attracts the N-pole, [1] S-pole then turns towards **X** until adjacent to **X** when magnet completes the 180° rotation [1]

4 Draw two straight lines crossing at middle, place plotting compass directly above where lines meet, turn paper so plotting compass points along one line, [1] place bar magnets equidistant on opposite sides of plotting compass, [1] move one magnet along the line so plotting compass points directly along the perpendicular line, magnet that is further from plotting compass must be stronger than other magnet, [1] because its effect cancels out effect of other magnet at a greater distance [1] (**or** valid alternative methods described correctly [2] and explained [2])

P15.2

1 Increase number of coils, [1] increase current, [1] use iron core [1]

2a See P15.2, Figure 1: concentric circles round wire, [1] lines of force in anticlockwise direction [1]

b Plotting compass points in same direction as nearest field line [1]

3a Reverses direction plotting compass points [1]

b Gradually moves towards North, [1] magnetic field of current-carrying wire becomes weaker further from wire [1] so Earth's magnetic field has greater effect [1]

4a See P15.2, Figure 3: field lines are loops which pass through solenoid [1] and loop round outside, [1] lines in solenoid are parallel to solenoid axis [1]

b i Plotting compass points parallel to the axis [1] in a direction consistent with current direction in solenoid [1]

ii Plotting compass points turns more and more towards North [1] because field of solenoid becomes weaker [1] so Earth's magnetic field has more effect [1]

P15.3

1a Insulated wire wrapped in tight coils along iron bar, [1] coil connected to suitable voltage supply so current passes through coil and magnetises bar [1]

b Electromagnet can be switched on and off [1] can increase the strength of electromagnet (if object heavier) [1]

2 **C, B, E, D, A** (1st 2 correct, [1] last two correct [1])

3a Line graph with axes correctly labelled, [1] suitable scales, [1] correct points, [1] best fit curve [1]

b 0.10 A current holds iron plate on the electromagnet, more than 0.10 A needed to hold any weight attached to iron plate, [1] the greater the current above 0.10 A the more weight can be held, [1] but increase of weight less for equal increases of current, [1] increasing current beyond 2.5 A unlikely to enable electromagnet to hold more than about 12 N [1]

4a Current through electromagnet coil magnetises core, [1] armature is pulled on to core which opens make-and-break switch and cuts current, [1] electromagnet loses its magnetism so make-and-break switch closes and cycle repeats [1] continually so armature vibrates [1]

b Buzzer has smaller mass than bell [1] so buzzer responds faster to switching on and off of electromagnet [1] which enables make-and-break switch to switch on and off at faster rate, so buzzer vibrates at higher frequency [1]

P15.4

1 Any two from: the size of the current, [1] strength of magnetic fi eld, [1] the angle of the conductor to the magnetic field [1]

2a Any three from: when current passes through coil of motor, a force acts on each side of coil due to magnetic field of magnet in motor, [1] force on each side has turning effect on coil and because current on each side in opposite directions, forces also in opposite directions so motor turns/moments in same direction on coil, [1] each half turn the commutator swaps from one brush to the other, so current round coil reverses direction, [1] without split-ring commutator, forces would reverse and coil would turn back, vibrating, [1] action of split-ring commutator allows forces to continue to turn coil in one direction [1]

b Current in opposite direction [1] so force on each side in opposite direction, coil rotates in opposite direction [1]

3 **A: up [1]**
B: up [1]

4 $\frac{0.024\,N}{1.8\,A \times 0.035\,m}$ [1] = 0.38 T [1]

P15.5

1a P.d. induced in wire when it cuts magnetic field lines, [1] this causes current to pass through wire and ammeter while wire is cutting field lines [1]

2a No deflection of pointer [1]

b Larger deflection [1]

c Larger deflection [1]

3 Induced current always creates magnetic field that opposes movement of magnet, [1] if magnet is moving, direction of induced current creates a magnetic field which repels magnetic field of coil if magnet is moving into coil [1] or attracts magnet if moving out of coil [1]

4 As magnet approaches and enters coil, pointer deflects one way briefly, [1] then deflects briefly in opposite direction as magnet moved out and away from coil, [1] second deflection bigger than first [1] because magnet accelerates as it falls through tube so its speed on leaving coil is greater on average than when it was entering coil [1]

P15.6

1a Split-ring commutator reconnects coil opposite way round in circuit every half-turn, each time coil is perpendicular to magnetic field lines, [1] so induced p.d. does not reverse its polarity [1]

b Correct half-wave form [1] half-wave part and zero part equal [1]

2a P.d. induced in coil when sides of coil cut across field lines, [1] p.d. reverses direction each time coil is at position where its sides are moving parallel to the field lines, [1] this happens every half-turn of coil so one full turn of coil corresponds to one full cycle of a.c. [1]

b Potential difference has greater peak value (i.e. amplitude) [1] and time period is less (or greater frequency) [1]

3a Peak value smaller, [1] waves stretched more across screen [1]

b Sides of coil cut more slowly across field lines so p.d. at any position of coil less than when coil spins faster, [1] waves more stretched out across screen because time for each cycle longer [1]

4 No induced current in the coil so no magnetic field around the coil [1] so not opposing the movement of the coil in the magnetic field [1]

P15.7

1a To increase or decrease peak value (or amplitude) of an alternating voltage, [1] step-up transformer used to increase voltage, step-down transformer is used to decrease voltage [1]

b i If mains supply fails, battery takes over [1]

ii Steps down [1]

2a 4000-turn coil [1]

b Iron core is magnetised [1] changes in the magnetic field through the secondary coil would not be enough to induce a p.d. [1]

3a Two insulated coils of wire wound on an iron core, [1] primary coil connected to a.c. supply, alternating p.d. induced in secondary coil, [1] transformer used to change amplitude (**or** peak value) of alternating p.d. [1]

b A.c. passed through primary coil [1] creating an alternating magnetic field in iron core [1] changing magnetic field moves through secondary coil [1] so an alternating p.d. induced in secondary coil [1]

4a D.c. in primary coil does not produce alternating magnetic field [1] so no p.d. induced in secondary coil [1]

b Current would short-circuit across wires/through iron core instead of passing through them and cause coil to overheat or blow a fuse [1]

P15.8

1 There will be more coils on the primary (20 × more/1200) [1] as this is a step down transformer [1]

2a $n_P = \frac{240\,V \times 100}{12\,V}$ [1] = 2000 turns [1]

b i 3.0 A (= 36 W/12 V) [1]

ii 0.15 A (= 36 W/240 V) [1]

3a $n_S = \frac{6.0\,V \times 1150}{230\,V}$ [1] = 30 turns [1]

b $I_P = \frac{6\,V \times 1.0\,A}{230\,V}$ [1] = 0.026 A [1]

P16.1

1a Gravity [1]

b Milky Way galaxy [1]

c Gravity [1] outwards force from fusion reactions [1]

2a i Jupiter [1]

ii Mercury [1]

b Planets nearer to Sun than Earth too hot for liquid water to exist on their surface, [1] planets further than Earth too cold for liquid water to exist [1]

3a Particles in clouds of dust and gas pulled together by gravitational attraction, [1] clouds became increasingly concentrated to form a protostar, [1] as protostar became denser its particles collided more and its temperature increased [1] until it became hot enough for nuclei of hydrogen atoms to fuse together, [1] forming helium nuclei and releasing enough energy to make protostar emit light [1]

b Any three from: star in which hydrogen nuclei in core fuse to form helium nuclei, [1] gravitational collapse is prevented by force from nuclear reaction [1] forces in equilibrium, [1] Sun is a main sequence star because most of its core consists of hydrogen nuclei [1]

4 Earth would be frozen if far from Sun [1] because it would receive much less energy from Sun, [1] if close to Sun, water on its surface would evaporate [1] and atmosphere would be lost as heated so much it its particles would leave Earth and go into space [1]

P16.2

1a **B**, **D**, **C**, **A** [1]

b i Main sequence [1]

ii Fade out, go cold, and become black dwarf [1]

2a i Gravitational attraction acting on its mass [1]

ii Fusion reactions outwards [1]

b White dwarf cools down, [1] when it no longer emits light it can no longer be seen it is a black dwarf [1]

3a i Hydrogen [1]

ii Uranium [1]

iii Helium, iron [1]

b Any two from: red giant star much larger than neutron star, [1] neutron star consists only of neutrons whereas red giant core contains helium and other light elements, [1] red giant star emits light and neutron star does not [1]

4a Sun and Solar System formed from debris of a supernova, [1] most uranium formed from the supernova still exists because uranium has half-life comparable with age of Earth [1]

b Some plutonium-239 was created in the supernova from which Sun and Solar System formed, [1] about 5000 million years ago, which is many half-lives of plutonium-239 so the plutonium-239 that was created has decayed into other elements [1]

P16.3

1a Force of gravity [1]

b i Towards centre of the Earth [1]

ii Towards centre of the circle [1]

c Difference: satellite orbits a planet and planets orbit the Sun [1] Similarity: Circular or almost circular orbits [1]

2 Satellite accelerates towards the Earth, [1] acceleration causes a change in direction, [1] change in direction causes velocity to change [1]

3a Geostationary satellite orbits once every 24 hours, [1] GPS satellite orbits in half this time, [1] the larger the orbit, the longer it takes so GPS satellite in lower orbit [1]

b Weather satellite takes less time to orbit than GPS satellite, [1] the larger the orbit, the longer it takes so weather satellite in lower orbit than GPS satellite [1]

4a Jupiter is slowest, [1] it is about 5 times further from the Sun than Earth because light takes 5 times longer to reach it, [1] but takes about 11 times longer to go round its orbit so its speed must be less than Earth, [1] Jupiter is about 13 times further than Mercury (as light from the Sun takes about 13 times longer to reach it) [1] but Jupiter takes about 120 times as long to orbit the Sun as Mercury , so must be slower than Mercury [1] (**or** for a circular orbit, speed of a satellite = circumference ÷ time it takes to go round once, [1] circumference of a circular orbit is proportional to its radius [1] so speed of a satellite proportional to orbit radius ÷ time satellite takes to go round once, [1] this ratio is equal to 1.6 for Mercury, 1 for Earth and 0.4 for Jupiter, [1] so Jupiter is slowest because it has the lowest ratio [1])

b Mercury is fastest: takes around 0.25 years to go round its orbit although its orbit is 0.39 times size of Earth's, [1] so Mercury is faster than Earth and Jupiter [1] (**or** Mercury has highest ratio in the analysis in **a** [1] so it is the fastest [1])

P16.4

1a i Receding [1]

ii Approaching [1]

b Blue-shift means Andromeda must be moving towards us [1]

2a Earth, Sun, Andromeda galaxy, universe [1]

b i They have red-shifts of same order of magnitude as distant galaxies [1]

ii As brightness of a quasar seen from Earth is same as brightness of billions of stars in a distant galaxy even though a quasar is much smaller than a galaxy, [1] power output of a quasar is about same as billions of stars/a galaxy [1]

3a Light from a light source (e.g. galaxy) moving away from us has increased wavelength due to motion of source, [1] this increase in wavelength called red-shift [1]

b i **Y** [1]

ii **X** [1]

4 Any three from: these galaxies have different speeds because they have different red- or blue-shifts, [1] ones with red-shifts are moving away from us, [1] ones with blue-shifts are moving towards us, [1] some galaxies may be moving across but we can't tell because they are so far away [1]

P16.5

1a Universe began as a small, hot dense region in space which has been expanding ever since [1]

b It provided evidence that universe was created in a massive explosion [1]

2 **C D A B** [1]

3a Distant galaxies are accelerating away from each other [1]

b Universe would stop expanding [1] and go into reverse, ending in Big Crunch [1]

4a $\approx$ (150 000 km/s ÷ 22 km/s) × 10^6 light years [1] $\approx 6.8 \times 10^9$ **or** 6800 million light years [1]

b $\frac{6.8 \times 10^9}{100000} \approx 70000:1$ [1]

MS1

1a Ratio = 5.6 g : 25.2 g = 1 : 4.5 [1]

b Width of strip = width of card / 4.5 [1] $= \frac{210\,\text{mm}}{4.5}$ = 47 mm [1]

2a $\frac{30}{24} = 1.25$, [1] ∴ ratio of car speed to coach speed = 1.25 : 1 [1]

b For the same time, $\frac{\text{distance travelled by car}}{\text{distance travelled by coach}} = 1.25$ [1]
∴ distance travelled by car = 1.36 × distance travelled by coach = 1.25 × 12 km = 16 km [1]

MS2

1a 1.679 s = 1.68 to 3 s.f. [1]

b 1.66 s [1]

c 1.61 s and 1.65 s [1]

2a 1 [1]

b 3 [1]

c 3 [1]

d 4 [1]

e 3 [1]

f 3 [1]

3a Gas was hardly used in 1990 whereas in 2010 it was used to generate about 40% of UK electricity, [1] coal was used to generate about two-thirds of UK electricity in 1990 but in 2010 this reduced to about one third of UK electricity, [1] oil generated less than 10% of UK electricity in 2010 but this is much more than in 1990, [1] nuclear power stations provided just under 20% of UK electricity in both 2010 and in 1990 [1]

b i Renewable energy sources in 2010 were used to generate less than 10% of UK electricity whereas the comparable figure for fossil fuels was about 70% [1]. Comparable figures for 1990 were about 5% for renewable energy and 75% for fossil fuels. [1]

ii Since 1990, fossil fuel use as a percentage of total UK electricity generation down by about 5% [1] and renewable energy use up by about same amount [1]

4 Speed ~ 0.1 m/s, [1] distance in 1 minute ~ 6 m [1]

MS3

1a i m [1]

ii m^3 [1]

iii kg [1]

iv N [1]

b i 7.2×10^{-2} m [1]

ii 1.6×10^{-5} m^3 [1]

iii 3.85×10^8 J [1]

iv 5.6×10^4 N [1]

2a The circumference of the Earth is much less than the circumference of the Sun [1]

b Atmospheric pressure is proportional to altitude [1]

c 4.5 kJ is about the same as 4.61 kJ [1]

3a $k = \frac{F}{e}$ [1]

b Unit of k is the unit of force (N) divided by the unit of distance (m) = N/m [1]

4a $e = \sqrt{\frac{2E}{k}}$ [1]

b $\sqrt{\text{Extension}}$

5a $S = \frac{v^2 - u^2}{2a}$ [1]

b 69 m [1]

MS4

1a i Gradient [1] **ii** y-intercept [1]

b i Positive gradient straight line [1] through origin [1]

ii Straight line below A with lower gradient and lower y-intercept [1]

2 $y \rightarrow F$, $x \rightarrow e$, $m \rightarrow k$ and $c \rightarrow y$-intercept [2]

3 $y \rightarrow v$, $x \rightarrow t$, $m \rightarrow a$ and $c \rightarrow u$ [2]

4 Gradient of tangent at 4 s [1] determined from gradient triangle [1] = 7.5 m/s (7.3 to 7.7 m/s acceptable) [1]

MS5

1a 4.4 m^2 [1]

b 0.21 m^2 [1] (both to 2 s.f.)

2a 3.4 m^3 [1]

b 14 m^2 [1] (both to 2 s.f.)

3a Area of each coloured triangle is same, taking base of each triangle as 1.6 m, height of each triangle is half width of kite = 0.575 m $\left(= \frac{1.15\,\text{m}}{2}\right)$, [1] area of each triangle = $\frac{1}{2}$ × height × base = 0.5 × 0.575 m × 1.60 m = 0.46 m^2 [1] so area of kite = area of both triangles = 2 × 0.46 m^2 = 0.92 m^2 [1]

Index

Appendix 1: Physics equations

Students should be able to remember and apply the following equations, using SI units, for their examinations.

Word equation	Symbol equation
kinetic energy = 0.5 × mass × (speed)^2	$E_k = \frac{1}{2} mv^2$
gravitational potential energy = mass × gravitational field strength × height	$E_p = mgh$
power = $\frac{\text{energy transferred}}{\text{time}}$	$P = \frac{E}{t}$
power = $\frac{\text{work done}}{\text{time}}$	$P = \frac{W}{t}$
efficiency = $\frac{\text{useful power output}}{\text{total power input}}$	
efficiency = $\frac{\text{useful output energy transfer}}{\text{total input energy transfer}}$	
charge flow = current × time	$Q = It$
potential difference = current × resistance	$V = IR$
power = potential difference × current	$P = VI$
power = (current)^2 × resistance	$P = I^2 R$
energy transferred = power × time	$E = Pt$
energy transferred = charge flow × potential difference	$E = QV$
density = $\frac{\text{mass}}{\text{volume}}$	$\rho = \frac{m}{V}$
weight = mass × gravitational field strength	$W = mg$
work done = force × distance (along the line of action of the force)	$W = Fs$
force applied to a spring = spring constant × extension	$F = ke$
distance travelled = speed × time	$s = vt$
acceleration = $\frac{\text{change in velocity}}{\text{time taken}}$	$a = \frac{\Delta v}{t}$
resultant force = mass × acceleration	$F = ma$
H momentum = mass × velocity	$p = mv$
wave speed = frequency × wavelength	$v = f\lambda$

The following equations are GCSE Physics only and will not be required for the GCSE Combined Science: Trilogy course.

GCSE Physics only

moment of a force = force × distance (normal to direction of force)	$M = Fd$
pressure = $\frac{\text{force normal to a surface}}{\text{area of that surface}}$	$p = \frac{F}{A}$

Students should be able to select and apply the following equations from the Physics equation sheet.

Word equation	Symbol equation
(final velocity)^2 – (initial velocity)^2 = 2 × acceleration × distance	$v^2 - u^2 = 2as$
elastic potential energy = 0.5 × spring constant × (extension)^2	$E_e = \frac{1}{2}ke^2$
change in thermal energy = mass × specific heat capacity × temperature change	$\Delta E = mc\Delta\theta$
period = $\frac{1}{\text{frequency}}$	$T = \frac{1}{f}$
H force on a conductor (at right angles to a magnetic field) carrying a current = magnetic flux density × current × length	$F = BIl$
thermal energy for a change of state = mass × specific latent heat	$E = mL$

The following equations are GCSE Physics only and will not be required for the GCSE Combined Science: Trilogy course.

GCSE Physics only

H pressure due to a column of liquid = height of column × density of liquid × gravitational field strength	$p = h\rho g$
H force = $\frac{\text{change in momentum}}{\text{time taken}}$	$F = \frac{m\Delta v}{\Delta t}$
magnification = $\frac{\text{image height}}{\text{object height}}$	
H $\frac{\text{potential difference across primary coil}}{\text{potential difference across secondary coil}} = \frac{\text{number of turns in primary coil}}{\text{number of turns in secondary coil}}$	$\frac{V_p}{V_s} = \frac{n_p}{n_s}$
H potential difference across primary coil × current in primary coil = potential difference across secondary coil × current in secondary coil	$V_p I_p = V_s I_s$
For gases: pressure × volume = constant	pV = constant

OXFORD
UNIVERSITY PRESS

Great Clarendon Street, Oxford, OX2 6DP, United Kingdom

Oxford University Press is a department of the University of Oxford. It furthers the University's objective of excellence in research, scholarship, and education by publishing worldwide. Oxford is a registered trade mark of Oxford University Press in the UK and in certain other countries.

First published in 2024

British Library Cataloguing in Publication Data
Data available

978-138-205151-4

10 9 8 7 6 5 4 3 2 1

The manufacturing process conforms to the environmental regulations of the country of origin.

Printed in Great Britain by Ashford Colour Press Ltd, Gosport.

Acknowledgements
I'd like to thank the awesome team at the OUP: Sophie Ladden and Thanucha Sooriapatham, for their constant support and encouragement; and my editors, for helping me appreciate 'old' physics in new ways, and for challenging me to reframe my ideas and language to strive towards authentic inclusion.

The publisher and authors would like to thank the following for permission to use copyright material:

Cover: Sua Balac/closer&closer.

Artwork by Q2A Media, Michal Bednarski, Erwin Haya, Aptara, Jeffrey Bowles, Wearset Ltd., Peter Bull Art Studio, and Oxford University Press.